Praise for *The Telomerase Revolution*

"*The Telomerase Revolution* is a remarkable book, telling a fascinating story that pulls together at last a single coherent theory of how and why growing old leads to so many different forms of illness. It also offers a tantalizing promise that we might soon know not only how to cure and prevent age-related diseases, but how to reset the aging process itself. Michael Fossel is a radical optimist."

—*Matt Ridley, author of* Genome *and* The Rational Optimist

"*The Telomerase Revolution* breaks down centuries of human thought on aging and uproots outdated ideologies that have led to nothing but worthless snake oil products. Dr. Fossel's exciting book is opening doors to extended healthspan that can change human history, and it's all grounded in solid scientific research."

—*Noel Patton, founder and chairman of T.A. Sciences*

"Michael Fossel's compelling argument for the telomere approach to reversing aging isn't just worth a look—it's like reading the words of Virgil as he leads us along the mysteries of aging."

—*Alexey Olovnikov, PhD, Institute of Biochemical Physics and Russian Academy of Sciences*

"Dr. Fossel has made a superb case for his belief that telomeres and telomerase play an essential role in the biology of aging both in humans and in other animals. His views were once in the minority, but more recent advances in how these molecules work have made his present book a valuable contribution to our understanding of the fundamental biology of aging. Adding to its value is that it is clearly written and well organized."

—*Leonard Hayflick, PhD, Professor of Anatomy, University of California, San Francisco*

"Aging is not an irreversible degenerative process, but an epigenetically determined physiological mechanism, which must not be confused with age-related diseases caused by lifestyle choices. Here, we have an effective and clear guide to understanding how we get old and how to tame aging in a few years."

—*Giacinto Libertini, MD, member of the Italian Society of Evolutionary Biology*

THE
TELOMERASE
REVOLUTION

THE TELOMERASE REVOLUTION

The Enzyme That Holds the Key to
Human Aging . . . and Will Soon
Lead to Longer, Healthier Lives

MICHAEL FOSSEL, MD, PHD

BenBella Books, Inc.
Dallas, Texas

BenBella Books, Inc.
10300 N. Central Expressway
Suite #530
Dallas, TX 75231
www.benbellabooks.com

Send feedback to feedback@benbellabooks.com

Printed in the United States of America
10 9 8 7 6 5 4 3 2 1

Library of Congress Cataloging-in-Publication Data:
Fossel, Michael.
   The telomerase revolution : the enzyme that holds the key to human aging, and will soon lead to longer, healthier lives / Michael Fossel.
      pages cm
   Includes bibliographical references and index.
   ISBN 978-1-941631-69-0 (hardback) — ISBN 978-1-941631-70-6 (electronic) 1. Aging—Molecular aspects. 2. Telomerase. I. Title.
   QP86.F69 2015
   612.6'7—dc23
                                2015026608

Editing by Erin Kelley
Copyediting by Eric Wechter
Proofreading by Jenny Bridges and
   Lisa Story
Indexing by Clive Pyne, Book Indexing
   Services
Text design by Publishers' Design and
   Production Services, Inc.

Artwork by Aaron Edmiston
Text composition by PerfecType,
   Nashville, TN
Cover photo © Eraxion www.
   fotosearch.com
Cover design by Brian Barth
Jacket design by Sarah Dombrowsky
Printed by Lake Book Manufacturing

Distributed by Perseus Distribution
www.perseusdistribution.com
To place orders through Perseus Distribution:
Tel: (800) 343-4499
Fax: (800) 351-5073
E-mail: orderentry@perseusbooks.com

Significant discounts for bulk sales are available. Please contact Glenn
Yeffeth at glenn@benbellabooks.com or (214) 750-3628.

To those with minds open to logic and eyes open to data:
May others be as open to you as you are
to the world around you.

To those who, aging and suffering,
hear others tell you nothing can be done:
They're wrong.

Contents

⌇

Telomere Theory of Aging Timeline

1665: Robert Hooke discovers that organisms are made up of cells.

1889: Charles-Édouard Brown-Séquard, a pioneer in endocrinology, claims that injected extracts of animal testis tissue (guinea pigs, dogs, monkeys) rejuvenates humans and prolongs life.

1917: Alexis Carrel begins thirty-four-year in vitro experiment with chicken-heart cells, apparently showing that individual cells are immortal. Carrel's research becomes a scientific paradigm until it is disproven in 1961.

1930s: Serge Voronoff implants testes and ovaries of chimpanzees and monkeys in humans as anti-aging therapy.

1934: Mary Crowell and Clive McCay of Cornell University double the life expectancy of laboratory rats through severe calorie restriction. To date, this has not been definitively duplicated in humans or other primates.

1938: Hermann Muller discovers the telomere, a structure at the ends of chromosomes.

1940:   Barbara McClintock describes telomeres' function as protecting the ends of chromosomes. She later wins the Nobel Prize.

1961:   Leonard Hayflick exposes the procedural error in Carrel's experiment and introduces the concept of the Hayflick Limit, which shows that the cells of any given multicellular species divide a limited number of times before they become aged and dysfunctional (e.g., forty times in human fibroblasts).

1971:   Russian scientist Alexey Olovnikov publishes a hypothesis that telomere shortening is the mechanism responsible for the Hayflick Limit.

1972:   Denham Harmon publishes mitochondrial free-radical theory of aging.

1990:   Michael West founds Geron Corporation with the initial goal of finding a way to intervene in the aging process based on telomere research.

1992:   Calvin Harley and his colleagues discover that patients with Hutchinson-Gilford progeria, a genetic disease in which children die of "old age" by the age of 13, are born with short telomeres.

1993:   Michael Fossel begins work, based on Geron's research, on the first book about the developing understanding of how and why aging occurs. *Reversing Human Aging* is published in 1996.

1997–1998: First peer-reviewed articles appear in the *Journal of the American Medical Association* suggesting that telomerase might be used to treat age-related diseases, authored by Michael Fossel.

1999:   Geron demonstrates that telomere shortening is not only related to cell aging but causes it, and that re-lengthening telomeres resets aging in cells.

2000:   Geron patents the use of astragalosides for use as telomerase activators.

Early 2000s: Geron and other research laboratories show that lengthening telomeres reverses aging not only in cells but in human tissues. Rita Effros conducts research at UCLA on immune aging and telomerase activators.

2002:   Geron shelves pharmaceutical development of telomerase activators to concentrate on cancer therapies, sells nutriceutical rights for astragalosides to TA Sciences.

2003:   Sierra Sciences founded, begins research on screening potential telomerase activators.

2004:   Oxford University Press publishes the textbook *Cells, Aging, and Human Disease* by Michael Fossel.

2005:   Phoenix Biomolecular begins research on a new technology to deliver telomerase directly to cells. Insufficient funding brings the project to a premature end.

2006:   TA Sciences markets first nutriceutical telomerase activator, TA-65, derived from the plant *Astragalus membranaceus.*

2007:   First human trials of a telomerase activator begin, as TA Sciences begins to collect data on users of TA-65.

2009:   Nobel Prize awarded to Elizabeth Blackburn, Carol Greider, and Jack Szostak for their academic research on telomerase.

Early 2010s: First companies founded to assess aging and the risk of disease by measuring telomere lengths: Telomere Diagnostics (founded by Cal Harley, formerly of Geron, in Menlo Park, California) and Life Length (founded by Maria Blasco in Madrid, Spain).

2011:   Ron DePinho, then at Harvard, shows that aging can be reversed in certain genetically modified animals.

2011:   Geron sells rights to *all* their telomerase activators to TA Sciences.

2012:   Maria Blasco at the Spanish National Cancer Research Centre in Madrid reverses many aspects of aging in several animal species.

2015:   Telocyte, the first biotech company dedicated to using telomerase genes to cure Alzheimer's disease, founded.

Introduction

In recent years, scientists have made extraordinary progress in understanding human aging. This research now brings us to the cusp of a real medical breakthrough—the ability to slow and even reverse the aging process and to cure a wide variety of age-related diseases.

You are right to be skeptical. Charlatans and dreamers—not to mention cosmetics companies—have been promising a cure for aging for centuries. The challenge is enormous, of course, and we are still just at the beginning.

But we now have a fairly clear understanding of the basis of human aging, which we'll explore in detail in this book. Based on that understanding, we also have some early therapies that have shown some modest results in changing the aging process. And we are close to human testing of therapies with considerably more promise.

Much of this research has gone unnoticed by the general public. In this book, I'll lay out the incredible breakthroughs that have been achieved so far and what we are on the verge of accomplishing. This has required a paradigm shift in the way aging is understood. As always, old paradigms die slowly, often frustratingly so.

As a doctor, my emphasis has always been on clinical results. Understanding the nature of aging is essential, of course. But the goal isn't simply to achieve understanding. The goal is to develop techniques to extend lives, cure diseases, and reduce suffering.

To accomplish this requires not just fundamental research, but the will of corporate boards who control the funding required for drug development and testing. I'll also share with you some of the inside stories of the often challenging process of making progress in a field with shifting corporate priorities and outdated paradigms.

I have been involved in the field of aging for more than thirty years, both as a clinician and as a scientific researcher. I've devoted my career to understanding the underlying causes of aging and developing therapies that have the potential to change the aging process. I've also devoted considerable time to getting my scientific peers to understand the latest developments in the field, both as editor of the *Journal of Anti-Aging Medicine* and as author of the textbook *Cells, Aging, and Human Disease* (Oxford University Press).

This book is my attempt to bring the latest research on aging to the general public. I think you'll find it enlightening, surprising, and ultimately quite hopeful.

CHAPTER ONE

Theories of Aging

I don't want to achieve immortality through my work. I want to achieve it through not dying.

— Woody Allen

Around 70,000 years ago, the first human beings—our direct ancestors—faced competition from Neanderthals and Homo erectus. These competitors were strong, intelligent, and fully capable of both language and tool-making. We were relatively slight and had little to recommend us as survivors as we moved into direct rivalry with earlier hominids. Our single major advantage was an odd one, an advantage that at first sight might seem to be a disadvantage. We were able to think and talk about things that don't actually exist.

This made all the difference.

These were abstractions like *tomorrow, god, art, science, dreams,* and *compassion*. You can't throw a spear at these things or eat them, steal them, break them, or destroy them. Yet these things not only made us human, but, oddly enough, made us far better survivors. Not only could we discuss intangible things that were necessary to social organization—like loyalty, cooperation, and strategy—but we could imagine things that that could be *made*—like weapons, tools, agriculture, and laws.

These abilities—abstract thought and imagination—are the foundation of our ability to *create*. Humans create not only art and tools, but also theories—religious and scientific explanations of how the world works—which ultimately allow us to change our own reality. Scientific advancement directly depends on this skill. We construct a vision of how reality works, we test our explanation, and then we use it to improve reality. A scientific theory is just that: a vision of reality that we can test and then use to improve our world. We cure disease, we grow food, and we gradually make human life easier and safer.

Man is the only creature that can do this. This ability to work with abstract concepts is lacking in other animals, even our closest relatives, chimpanzees and gorillas.

The key to using a theory to improve human life—or turning a dream into reality—is to have the right tools and the knowledge to use them. I often think of it as having a ship and a map.

OUTSMARTING A GORILLA

Koko was the first gorilla to use sign language. When she was three years old, I became her babysitter for six hours every week for a year. Koko understood more than a thousand signs and was adept at inventing games. She had learned to stop biting me (only after I bit her back) but would pull my laundry bag over her head and body—leaving nothing but two black, furry legs sticking out from the bottom of my gray cloth bag—then leap at me from the kitchen counter and try to chase me down. Her "rule" was that if she could catch me, she could bite me—but only as long as she kept the laundry bag over her head so I couldn't see her biting. Somehow, a gray laundry bag made all the difference. It let her create a new way to play with me. On the other hand, while she was clearly smarter than any other animal I have ever met, she never mastered signs for the abstract concepts that are central to both human thinking and human society.

Sometimes the ship is simple, but the map is complex. To prevent smallpox, the ship can be as simple as a sharp needle infected with cowpox. This is all we need to vaccinate against smallpox, if we know how. But first we needed the map; we needed to know about germs, vaccination, smallpox versus cowpox, infections, and so on.

This chapter discusses the maps we have drawn as we tried to understand aging. As we will see, there has been no single consensus map, but rather a myriad of diverse maps and clashing interpretations of those maps. Now we are beginning to coalesce around a map that genuinely explains aging. As for the ship—the tools we need to change aging have become more sophisticated over the past 500 years, until, as of the last decade or so, we are at the cusp of clinical breakthroughs.

Let's begin by understanding the competing maps we've drawn to explain aging. They all contain an element of truth, but none fully solves the riddle.

The Entropic Theory of Aging

At first, it wasn't clear that aging was even a problem to be solved. The aging of living things is hardly unique. Mountains age, galaxies age, even the universe itself ages. In fact, the second law of thermodynamics states that the entropy of any closed system always increases, that disorder always increases. That's why after a few years of being left alone, your car won't start. After a few million years, a mountain range is reduced to dust. And after 11 billion years or so, the sun itself will grow cold. Everything ages.

Life depends on order, structure, and organization. With too much disorder, life cannot maintain itself. And so the mystery appeared solved. Organisms age because the very nature of the physical universe requires it.

A number of specific theories fall under the general heading of entropic explanations of aging. These theories suggest that this basic fact of life—wear and tear—is sufficient to explain the aging process.

Many of these approaches are variations on a theme. The cross-linking theory suggests that all aging is due to molecules becoming linked over time, interfering with their normal function. A similar explanation blames advanced glycation end-products (AGEs) for dysfunction, as glucose molecules bind to protein molecules, causing an accumulation of these waste products and loss of function.

There are a host of other explanations that blame aging on the accumulation of various other waste products, such as lipofuscin, a pigmented lipid product that accumulates in many aging cells.

One particularly tempting variation focuses not on damage to the routine molecules and enzymes, but on the most critical set of molecules in living cells, the DNA. These theories posit that, over time, DNA slowly accumulates damage, reducing the ability to produce critical proteins. As the cell becomes more and more dysfunctional, aging ensues, and the cell finally fails altogether.

All of these theories are based on a fundamental truth: As time goes on, damage occurs. Molecules become linked, waste products are generated, and DNA is damaged. But these theories underestimate the incredible power of cellular regeneration. While it is true that some cells age and fall into disrepair, others remain in full health, living and reproducing without limit, despite cosmic rays, waste accumulation, and a changing environment.

For billions of years, all life was single-cellular, and these individual cells could reproduce indefinitely. Whether these cells aged in some ways is open to debate, but it's clear that with each reproductive cycle, with each splitting of an older cell into two daughter cells, the clock restarted. Each daughter cell was young and healthy.[1]

Life repairs and replaces its components at an amazing rate. If every part in your car were replaced each year, theoretically it could run forever. As we shall see, single-celled organisms do exactly that. This doesn't violate the law of entropy, because the Earth is not a closed system. Earth is constantly bathed in light and energy from the sun. The sun's nuclear fusion generates a tremendous

---

[1] Some single-celled organisms divide asymmetrically, with one daughter cell being damage-free and the other having some residue of damage. But this doesn't negate the main point, which is that single-celled organisms have thrived without aging for billions of years.

rate of entropy, but life uses the solar energy to maintain itself, so that it continues to flourish. There is no physical law that says an organism can't continue to live and thrive indefinitely, at least as long as the sun shines.

In summary, there is an entire category of theories that try to ascribe aging to entropy, explaining aging in terms of wear and tear, damage, and waste products. Although these theories contain a germ of truth, they don't offer a complete explanation. Some cells and organisms do succumb to entropy, but others do not. A deeper level of insight is required.

THE JELLYFISH AND IMMORTALITY

The ability to thrive and stay healthy indefinitely can extend beyond single-celled organisms. The *Turritopsis dohrnii*, known today as the "immortal jellyfish," apparently has the ability to reverse aging. This invertebrate reverses its aging process until it reaches the protozoan stage. Indeed, it is often called the Benjamin Button jellyfish. Unlike Benjamin Button, however, this jellyfish then begins aging again, repeating this process into, as far as we can see, infinity.

As the authors of a 1996 paper on the phenomenon stated, this reveals "a transformation potential unparalleled in the animal kingdom."[1] A later article in the *New York Times* said the finding "appeared to debunk the most fundamental law of the natural world—you are born, and then you die."[2]

Animals of the genus hydra also appear not to be senescent. Lobsters, while certainly not immortal, appear to grow and increase in fertility as they age, avoiding the symptoms of senescence that affect most multicellular life.

The jellyfish and the hydra strike yet another blow to the entropic theory of aging.

[1] Piraino, S., Boero, F., Aeschbach, B., et al. "Reversing the Life Cycle: Medusae Transforming into Polyps and Cell Transdifferentiation in Turritopsis Nutricula (Cnidaria, Hydrozoa)." *The Biological Bulletin* 190, no. 3 (1996): 302–12.
[2] Rich, N., "Can a Jellyfish Unlock the Secret of Immortality?" *New York Times*, November 28, 2012.

The Vitalist Theory of Aging

The notion that aging occurs because we "run out of something" is an old one. Centuries ago, it was called vitalism, and the idea can even be found in writings of the early Greeks, including Aristotle, Hippocrates, and Galen. We age because something in us—the vital spark that gives life—only lasts so long, and then we die because it has run out, leaving us no more than inanimate matter.

Generically, these sorts of explanations are called "rate of living" hypotheses. The most obvious of these explanations was the "heartbeat hypothesis"—that every living creature has a limited number of heartbeats. As you approach that critical value, you age; when you reach that value, you die. This offered a partial explanation for one of the most obvious of aging anomalies: Not every organism ages at the same rate. The thought was that because smaller animals have a more rapid heart rate (or metabolic rate or breathing rate), they age faster than larger animals. In this view, dogs age faster than humans because their hearts beat faster.

Variously called the life force, the *élan vital*, the vital spark, or simply the soul, this entire concept had been all but abandoned by science in the early twentieth century, because of failures of logic (does a cell have a heartbeat?) and lack of empirical support. But I discuss it here because this general idea, that aging is the result of something running out or running down, is still with us, albeit in modern form.

The central fallacy of ascribing aging to the loss of some critical component—whether a heartbeat, mitochondria, or a hormone—is that we immediately ask what causes aging within that component. If aging is caused by mitochondrial changes over time, then what causes those changes? If aging is caused by having only a fixed number of heartbeats, then what fixes that particular number? If aging were caused by the loss of a key endocrine gland, then what causes that endocrine gland to age?

The Hormonal Theory of Aging

The notion that hormone deficiencies cause aging is still quite popular. The earliest work can be traced to Chinese medicine. In Western medicine the field of endocrinology—the diagnosis and treatment of hormone-related diseases—blossomed in the 1800s. Endocrinology quickly became both mainstream science and accepted clinical medicine. As with many medical advances, however, this was rapidly followed by unfounded claims and wishful thinking.

The most spectacular claims centered around aging in the area of sexuality. These claims involved the use of the testicles (and more rarely, the ovaries) of young animals, which were variously eaten by, transplanted to, or extracted and injected into patients. The most prominent leader in the new field of endocrinology was Charles-Édouard Brown-Séquard, a world-famous physician who practiced in France, England, and the United States in the mid-1800s. He claimed that he "rejuvenated sexual prowess after eating extracts of monkey testis." Those who adhere to Mark Twain's suggestion that you should eat a live frog for breakfast, because nothing worse can then happen to your day, clearly haven't reckoned with Brown-Séquard's approach to self-improvement!

Truth being stranger than fiction, this approach to anti-aging therapy continued with the transplantation of chimpanzee testicles to human males (and monkey ovaries into human females). Performed worldwide by Serge Voronoff, this became the therapy craze of the 1930s and was so popular that monkey hunting was banned by the French government in their colonies, prompting Voronoff to try breeding monkeys for this purpose alone. Similar interventions became widespread in the United States, using both colored-water injections and goat testicle transplants.

Presently, there is still widespread belief that testosterone or estrogen can actually reverse the aging process. To a degree, this belief springs from the observation that our levels of such steroids fall with age. In most men, this fall is gradual; in most women, it occurs more observably at menopause.

THE VALUE OF GROWTH HORMONES

At an aging conference in Morocco, I was asked if there is any value in using growth hormone to treat aging. "Yes, of course," I replied. "There is a considerable value, although not in buying growth hormone, but in selling it. It doesn't do anything for aging, but there is certainly a market for it." The pharmaceutical firm, which sold growth hormone, did not invite me back.

This common assumption—if hormone levels decline with age, then hormone replacement will make me young again—is not only bad logic, but is contradicted by the medical data. Claims that hormone replacement therapy (HRT) makes some people feel younger are the same as claims made a century ago by those who used monkey testicles, rhino horn, and colored water.

Do hormones sometimes have therapeutic benefits? Yes.

Can hormones ever slow, stop, or reverse aging? No.

The Mitochondrial or Free-Radical Theory of Aging

Perhaps the most publicly well-known explanation of aging is the mitochondrial free-radical theory first published by Denham Harmon in 1972. Free radicals occur naturally, the side effect of metabolism, particularly the metabolism that happens within our mitochondria. As you may remember from high school biology, mitochondria are the "powerhouse" of the cell. Like powerful nuclear reactors, the mitochondria generate large amounts of energy. And, as with nuclear reactors, there's a considerable amount of waste.

As we burn metabolic fuels (such as glucose), our bodies create free radicals, charged molecules that disrupt other molecules. Fortunately, the overwhelming majority of free radicals are created inside our mitochondria and remain there, away from most of the important molecules in our cells and even further away from the DNA of our genes, which are hidden safely inside the cell's nucleus.

But those free radicals that escape wreak havoc on the complex biological molecules in our cells such as DNA, membrane lipids, and crucial enzymes.

Free-radical theory has a great deal of credibility. Some of the most important changes that occur in aging cells can be laid directly at the door of free radicals and the damage they cause within our cells. As our cells age, there are four important changes that occur with regard to free radicals: production, sequestration, scavenging, and repair.

The first change is the increase in *production* of free radicals. Young mitochondria produce few free radicals and a lot of energy. Old cells, however, have a higher ratio of free-radical production to energy production. And as more free radicals are produced, more damage occurs.

The second change—to *sequestration*—is that more free radicals escape from the mitochondria into the rest of the cell, even into the nucleus. This occurs because the lipid membranes that make up the walls of the mitochondria become leakier with age.

THE FATHER OF FREE-RADICAL THEORY

Denham Harmon, both "the father of free-radical theory" and the first proponent of the mitochondrial theory of aging, was a remarkable human being. (Sadly, he passed away in November of 2014.) Born almost a century ago, he finished a PhD, got interested in the causes of aging, went back to get his MD from Stanford, and then spent the rest of his life as a professor of medicine, trying to understand and explain human aging. In 1970, he helped found the American Aging Association (AGE). In 1985, he founded the International Association of Biomedical Gerontology (IABG). In working with Denham on the boards of both AGE and IABG, I have seen him listen to the thoughts of others around him for hours—often those with less knowledge or wisdom—politely and patiently. A man without hubris, he was thoughtful, kind, well-respected, and even revered by those in the aging community.

The third change affects *scavenging*. In young cells, free-radical scavengers effectively capture free radicals. Older cells produce fewer scavengers, so that more free radicals remain, inflicting greater damage.

The fourth change is that older cells are less able to *repair* free-radical damage. So, not only does the aging cell incur more free-radical damage—due to increased production and reduced sequestration and scavenging—but also the aging cell is slower to repair the damage. (In the case of damaged DNA, the rate of repair goes down; in the case of all other molecules, the rate of replacement goes down.)

These processes create a vicious cycle. All four of these processes—production, sequestration, scavenging, and repair—are interlinked, with the result that aging cells become increasingly dysfunctional at all levels.

Although it's tempting to see this avalanche of metabolic damage as the *cause* of aging, the conclusion that free-radical theory can explain aging is unwarranted. Free-radical theory has a certain elegance and an overwhelming acceptance among the public, but it also suffers from a major problem: It explains much of what happens as a cell ages, but it doesn't explain what *causes* these changes to occur. *Why* do these four processes—production, sequestration, trapping, and repair—change as we age? What starts the avalanche going downhill in the first place?

Some cells, for example human germ cells (sex cells), show none of these changes despite an unbroken line of ancestry going back for several billion years of life. So how is it that free radicals irreparably damage some cells within years, yet have no effect at all on germ cells or on single-cell organisms over billions of years?

Moreover, the elimination of free radicals, even if such a thing were possible, would be disastrous. We *need* free radicals to survive, as we use them to both modulate gene expression and kill microbes. If we lower the concentration of free radicals in healthy cells, the pattern of gene expression changes, and the cell becomes less functional. Our immune systems use high concentrations of free radicals to attack invading organisms such as bacterial infections. Free radicals may well be a driving force in the aging process, but they are also a normal and beneficial part of our physiological function.

When we try to intervene in aging by altering free radicals, the results are ambiguous at best. There is a valid body of work that suggests that we can increase the mean lifespan of some laboratory animals by minimizing free-radical damage, but there is no evidence that we can change a species' maximum lifespan, no matter what we do to free radicals.

Incidentally, a similar argument applies to discussions about oxidants and antioxidants. Living organisms require oxidation as part of the metabolic process. Oxidation is the process by which oxygen reacts with molecules to form carbon dioxide and water, releasing energy in the process. There's a tendency to believe that oxidation is another cause of aging, but the reality is more complex. Not only can we not survive without oxidation (and oxygen!), but there is no evidence that antioxidants have any effect on the aging process either. As with free radicals, too much uncontrolled oxidation can certainly cause problems, but free-radical production and oxidation are necessary parts of our metabolism. And neither of them can really be said to drive aging.

We cannot say we have explained the aging process until that explanation can predict which mitochondria, cells, and organisms *will* undergo aging and which ones *won't*. The mitochondrial free-radical theory of aging has great descriptive power, but it isn't predictive.

The Nutritional Theory of Aging

It may be a bit of an exaggeration to state that there is a nutritional theory of aging, but there has been an enormous amount written on the topic of extending life through diet.

It's beyond the scope of this book to rebut the millions of words written on this topic, but I can give you the bottom line, based on the science to date: While there certainly is evidence that a poor diet can create disease and that a good diet can avert disease, there is no evidence that an optimal diet can prevent or reverse aging.

NUTRITIONAL HOAXES

History is full of stories of people who lived remarkably long lives because they ate the right foods. Marco Polo, for instance, encountered Indian yogis who claimed they lived 150 to 200 years eating only rice, milk, sulfur, and—in spectacular disregard for their health and our skepticism—mercury. It's never been clear whether the yogis were pulling Marco Polo's leg or he was pulling ours. Either way, it's only one of hundreds of historical examples where the claim of long life doesn't depend on special nutrition so much as our own inherent gullible optimism.

Aging is not a nutritional disease. It doesn't matter how much or how well we eat; no dietary manipulation can stop or reverse the aging process.

In 1934, however, Mary Crowell and Clive McCay of Cornell University found that they could double the life expectancy of laboratory rats through severe calorie restriction. Definitive data on humans or other primates has yet to be established, but there is reason to believe that significant caloric restriction has the potential of significantly extending human life. (And even if not, it'll certainly feel longer).

Even so, there is no evidence that caloric restriction can stop or reset the aging process. Many researchers believe that caloric restriction is not the "experimental group" at all, but is actually the "control group." They point out that animals (and humans) have evolved to thrive on a low-calorie diet. In a natural environment, calories are hard to come by. We evolved to get by without a lot of food, and now—modern society being what it is—we are burdened with a glut of food and are unable to control our own intake. From this point of view, the wonder is not that we might live longer if we ate less, but that we survive as well as we do on the fast food, poor nutrition, and abundance of empty calories typical in the diets of people in developed nations.

The Genetic Theory of Aging

In the latter half of the twentieth century it became fashionable to explain the world in genetic terms, almost exclusive of any other viewpoint. We've now come to accept the notion that specific genes cause almost everything from heart disease to Alzheimer's dementia and from osteoarthritis to aging itself. While genetic explanations can have great power, they must be invoked with great care. Too often, they aren't true.

Frequently, it is simply assumed that genes are the cause of all diseases, including aging. But there are two big problems with the notion of "aging genes."

The first problem is that most traits (e.g., height), diseases (e.g., atherosclerosis), and complex changes (e.g., aging), are not attributable to a gene or even to a small number of genes. Certainly there are genes that correlate with these things, but the notion that one or a few genes *cause* any particular complex outcome is only occasionally accurate and usually naïve. In the case of height, for example, we know that there are genes, environmental factors, and epigenetic factors that play roles in determining your final stature. (Epigenetic factors are inheritable traits that are not part of the DNA sequence.) There is no single "height gene" responsible for your stature.

The second problem is that genes are less important than gene *expression*—epigenetics. Our narrow focus on genes has blinded us to the overarching importance of this. In the early 1900s, for example, there were biologists who believed that your toes and your nose had entirely different genes. On the contrary, the genes for every part of your body are precisely the same. The difference between one cell type and another is not the genes, but the pattern of gene expression—the epigenetic pattern. There is no toe gene, only a toe pattern of gene expression. And a distinct pattern of gene expression is found in every single definable cell or tissue. It's much like having a single symphony orchestra that can play Mozart, the blues, or the Grateful Dead; the difference is not the instruments but the score. Oddly enough, the difference between toe and nose cells is also exactly the difference between a young cell and an old cell: They have the same genes, but the pattern of expression is different. The difference between my cells at age six and age sixty is not genetic, but epigenetic. So the hunt for "aging genes" is a futile endeavor.

And yet, "aging genes" are supposedly identified regularly and, apparently, in earnest, although with little insight and less understanding. Certainly there are specific genes or alleles[2] that are more common in some people who have shorter lifespans, and other genes or alleles that are more common in those with longer life spans, but calling these "aging genes" is misleading.

As we will see, this same confusion extends to age-related diseases. Every year, we gleefully identify another handful of genes that supposedly cause Alzheimer's dementia or atherosclerosis. Again and again, the data simply shows not causation but correlation, and a minor correlation at that. One gene is said to account for 1 percent of all Alzheimer's cases, another gene for an additional 2 percent, leaving us with a lot of cases still to be accounted for. Somehow, the implication is always that we will someday identify the genes behind the remaining 97 percent of Alzheimer's cases if only we put more money into the research. Unfortunately, finding the genes that *cause* Alzheimer's is like finding the genes that *cause* aging.

The problem is not that we lack funding or researchers, but that we lack a good solid understanding of the role of genes—and how patterns of gene expression change as we age—in the basic processes of both aging and age-related disease. In short, much like the story of the man who lost his keys on a dark street, we are forever searching under the street light simply because the light is better, even though we actually dropped them a block away in a dark alley. We look for aging genes because they are easy to identify, simple to explain, and more likely to get funding in today's scientific climate.

---

[2] An allele is an alternative form of a gene. In a gene for eye color, you might have a blue allele or a brown allele.

Unfortunately, when it comes to aging and age-related diseases, the real answers are not in our genes, but in the patterns of gene expression.

The Blind Men and the Elephant

We've looked at aging from several points of view—free radicals, mitochondria, nutrition, hormones, wear and tear, genetics, cell biology, and so forth—and each answer has been so different that it looked as though they couldn't possibly all be right.

The classic analogy of the blind men and the elephant is appropriate here. Six blind men are asked to describe an elephant. The man who feels a leg says the elephant is a column. The one who feels the tail says the elephant is like a rope. The one who feels the trunk says the elephant is like a snake. The one who feels the ear says the elephant is like a fan. The one who feels the ribs says the elephant is like a wall, and the one who feels the tusk says the elephant is like a pipe. Each gives an accurate description of his particular part of the elephant, but none of blind men comes close to describing an elephant.

While each of the various theories of aging I've just described was to some extent credible, they were all incomplete. None was quite capable of explaining the entire elephant. Like the blind men, our academics have given an accurate description of their particular parts of the aging process. Each of these theories is based on valid and accurate data. Yet none of us has been able to describe the entire aging process. We were honest, intelligent, and well-intentioned, yet none of us could fit all the data into a single, correct explanation of how aging works.

How could we put it all together into a complete "elephant?"

As a professor of medicine, my own viewpoint focused on intervention; was there a way to prevent or cure the diseases of aging? Perhaps if we could truly understand the aging process, we could intervene in Alzheimer's dementia, atherosclerosis, and all of the other age-related diseases that were part of my daily medical practice.

Since 1980, in addition to teaching courses on biology and aging, I have been working as a researcher and as a physician treating aging adults. Also, I have spent considerable time working with children with early-aging syndromes. Children with Hutchinson-Gilford progeria (or simply "progeria") die of what looks like old age, typically at about the age of thirteen. These children not only look old, but their cells are old. They die of what we think of as age-related diseases, most commonly strokes and heart attacks. It's one thing to know a seventy-year-old man who dies of a heart attack in his back yard while throwing a ball to his grandchild. It's quite another to know a seven-year-old child who *looks* seventy

AGED CHILDREN: THE TRAGEDY OF PROGERIA

In any given year, there are several dozen children with progeria in various parts of the world whom I know personally. Typically, the parents brought these children to a doctor when they noticed they weren't growing normally. Because the syndrome is rare and relatively unknown even to many pediatricians, the children were lucky that the clinician recognized the syndrome and brought it to our attention.

At the turn of the twenty-first century, we had nothing to offer these children or their parents except kindness and the knowledge that others shared their affliction and understood their tragedy. The parents could ask other parents how they handled the constant health problems and could talk with us about what we knew, which was far too little. The children especially looked forward to the time, once a year, when we would bring them all together from around the globe. It was the one time in their short lives when they looked like all the other children around them.

Oddly enough, progeric children often resemble no one so much as they resemble each other. In one case, a Vietnamese girl was much more clearly progeric than Asian in her facial structure. She looked more like the other children than like her parents. At our annual meetings, children who were bald, who had prominent frontal veins and arthritic joints were everywhere, playing and joking, happy to finally be—in some strange sense that we all understood—home at last.

and who dies of a heart attack while playing catch with his young mother. The incongruity of a child's dying of age-related disease creates a deep and lasting impression.

Telomeres are DNA structures at the ends of chromosomes that shorten with each cell division. In 1992 we discovered that progeric children were born with short telomeres. They have telomeres characteristic of people in their seventies. This and other findings made it clear that aging—in normal people, in progeric children, in cells, in other organisms—is closely related to telomeres. But we also knew that there were many other reasonable views of aging, well-supported by data. How could we reconcile our growing knowledge of telomeres and cell aging with these other views of how aging works?

The problem was one of perspective.

There were numerous theories and endless data, but there was always some data that simply didn't fit into a single, coherent picture of the aging process. It was as if we had a thousand parts of a complex machine and dozens of ideas for to how to put them all together, yet everyone who tried to put them together to make a single, functional device had a few extra parts left over. Worse yet, the machine itself *never actually worked*.

I had an epiphany in the early 1990s when I attended a conference on aging held in Lake Tahoe, California. I had intended to go, listen to the latest information, and incorporate it into an updated medical textbook on aging.

The points of view presented at the conference were overwhelmingly different. Not only were there lectures on free radicals, evolution, and other facets of the problem, but I spent much of my time "translating" for those around me. Researchers were unfamiliar with common medical terminology ("What's a non-steroidal anti-inflammatory medicine?"), while physicians were equally unfamiliar with common research terminology ("What's a Southern blot?"). Because I have a foot in both camps, it often fell to me to help explain disparate points of view. At times, it was as though my role was to explain to the blind man holding the elephant's leg why the blind man holding the tail was also right.

In the middle of the conference, Cal Harley, a cell biologist and now a friend, gave a talk on the latest work on telomeres and cell aging. He pointed out that when you knew the age of a cell and you measured the amount of telomere length that cell had lost, those two numbers lined up precisely. If you knew one, you knew the other.

In a matter of minutes, everything I knew as a professor of medicine, everything I taught in my classroom crystallized into an entirely new pattern. I began to see how all the viewpoints, no matter how seemingly disparate and contradictory, fit together into a single, well-defined picture.

I found myself staring straight at the whole elephant.

The more I thought about it, the more I found all the pieces fitting together. Instead of multiple theories, each with only part of the answer, I saw the outline of a single theory in which all of our data and viewpoints clearly explained how we age *and* where we might intervene. I began to see how we might test the theory to prove it correct or incorrect. And I saw how we might use this new understanding to go much further.

I began to see how we could cure the diseases of aging.

CHAPTER TWO

The Telomere Theory of Aging

The telomere theory of aging states that telomere-controlled cell aging results in aging in the entire organism. It could more accurately be called either the cell senescence theory of aging or the epigenetic theory of aging. The limited theory of aging—that telomeres control cell aging—is well-established, but the general telomere theory of aging—that cell aging causes aging itself—is still not universally accepted.

In the 1990s, when I first began talking about the telomere theory, I felt very much alone. I wish I could at least say the scientific community rose up to rebut this theory, but mostly they just ignored it.

However, as I write these words in early 2015, the telomere theory of aging has become the dominant one, although it is far from fully accepted by all scientists. I'd estimate that roughly half of the experts in the field accept it. Most promisingly, younger scientists are far more likely to consider the theory uncontroversial.

The telomere theory of aging has moved to prominence because it accomplishes five critical things:

1. It clearly explains the mechanism that drives the aging process at the cellular level over time.
2. It explains why some cells age and some do not.
3. It incorporates the proven elements of the other various theories of aging.
4. It can successfully address the various objections to the theory.
5. Perhaps most important, it provides a clear path to clinical interventions, taking us beyond theory to an actionable map for improving our health.

The Hayflick Limit and the Cellular Basis of Aging

In the first half of the twentieth century, the conventional wisdom was that cells were immortal, and that aging was something that happened between cells. No one had a good idea of what that "something" was, but the reasoning was sound. Because single-celled organisms apparently didn't age, but multicellular life clearly did, didn't that mean that something was happening between cells rather than within cells?

This belief was supported by the work of Alexis Carrel, whose work appeared to show that cells were immortal. Carrel was a highly respected but controversial French surgeon and biologist who won the Nobel Prize in medicine in 1912 for his work on vascular suturing techniques. Carrel was a devout Catholic who, in 1902, claimed to have witnessed a miraculous cure of a dying woman at Lourdes. This claim forced Carrel to leave France, because the anticlerical atmosphere of French academia made finding employment difficult. He wound up in Chicago's Hull Laboratory, where his work on vascular sutures and the transplantation of blood vessels and organs led to his Nobel Prize.

In 1912, Carrel undertook his famous chicken heart experiments. He grew cells from a chicken heart in his laboratory, adding nutrient broth every day and carefully measuring cell divisions. Month after month for thirty-four years, Carrel and his colleagues found no signs of cell aging. Their cells could seemingly divide forever, without slowing, without ceasing, without failing in any way. If he was right, then cells were indeed immortal.

Carrel's theory stood undisputed for decades. But it was wrong.

Only much later was it discovered that there was a grievous flaw in Carrel's experimental procedure. The daily nutrient broth inadvertently contained young heart cells. Of course, as long as Carrel kept adding young cells, the cell cultures thrived. But without the daily addition of young heart cells, Carrel's cells would have soon died off.

Although some dispute their honesty, Carrel and his research colleagues might have been unaware of their mistake. Unfortunately, their work had far-reaching effects on all of biology. Not only did an entire generation believe their erroneous results, but the mistake still taints and biases some parts of biological theory over a century later.

Carrel's mistake was revealed in the early 1960s by a professor of anatomy at the University of California at San Francisco, Leonard Hayflick. Hayflick and his colleagues attempted to replicate Carrel's work. No matter how hard Hayflick and his team tried, they could not create an immortal cell line. They soon realized Carrel's mistake. Hayflick's team, unlike Carrel's, was very careful not to introduce new cells into the culture. They found that cell lines uniformly aged after a fixed number of divisions, eventually becoming unable to divide any further.

With some trepidation—and to the intense disbelief of their scientific audience—they published their work. Anyone who tried to replicate their experiment, carefully excluding the addition of new cells, found the same results. Carrel was wrong. Cells age.

From the work of Hayflick and his team arose the concept of the Hayflick Limit. Put simply, the Hayflick Limit says that most cells can divide a fixed number of times (about forty to sixty times for most human cells) and that the rate of reproduction gradually slows until cells become quiescent and incapable of further division. In other words, cells don't age because of the passage of time; cell divisions cause the cell to age. Hayflick identified the cell nucleus as the critical component in cellular aging, controlling what he called a cellular "clock."

I'm pleased to claim Dr. Hayflick as a close friend of more than thirty years. Hayflick doesn't suffer fools lightly, but he is a deeply honest man and one of the bravest people I know. He is also one of the most remarkable scientists in history, singlehandedly over-turning more than fifty years of dogma about aging. It took fifteen years and much derision before Hayflick's theory was accepted. As Hayflick said in a 2011 interview published in *The Lancet*, "To torpedo a half-century-old belief is not easy, even in science."[1]

Interestingly, the Hayflick Limit is not the same for every species or every type of cell. There is a correlation between lifespan and the Hayflick Limit. However, this correlation is imprecise, more suggestive than definitive. Mice have a lifespan of three years and a Hayflick Limit of fifteen divisions, while the Galapagos turtle, which lives for 200 years, has a Hayflick Limit of around 110 divisions. Human fibroblasts have a Hayflick Limit of between forty and sixty divisions.[2]

The implication of the Hayflick Limit for cellular aging is profound. It strongly suggests that aging occurs *within* cells, not *between* cells. There is no mysterious substance or organism-wide dynamic that drives aging. This idea has empirical support from experiments, as well as our knowledge of human disease. Non-dividing cells show no sign of cell aging, whereas in cells that divide, regardless of the actual time that passes, it is the number of cell divisions that determines how "old" the cells are, not the passage of time.[3] Like many other cells[4] in our bodies, the vascular cells in our coronary arteries and the glial cells in our brain divide, lose telomere length, and show aging changes, and these are the cells that cause disease in the heart and in neurons in the brain. The cardiac muscle cells and the neurons in the brain don't age, but they depend on other cells that do age, and when these dividing

---

[1] Watts, G. "Leonard Hayflick and the Limits of Ageing." *The Lancet* 377, no. 9783 (2011): 2075.

[2] Actually, the Hayflick Limit depends on which type of cell we look at. In the cases cited here, we are looking at the Hayflick Limit for fibroblasts, a typical cell in almost any species.

[3] Hayflick, L. "When Does Aging Begin?" *Research on Aging* 6, no. 99 (1984): 103.

[4] Takubo, K. et al. "Telomere Lengths Are Characteristic in Each Human Individual." *Experimental Gerontology* 37 (2002): 523–31.

cells age, the result is disease. Aging occurs within cells that divide, causing disease in other cells that may not divide—or age—at all.

Cell aging is well-accepted, but the more general model—that cell aging causes age-related diseases and aging of the body itself—has also become more accepted over time. If your cells are young, you are young. If your cells are old, you are old. Aging is a product of cellular senescence. It's that simple—and that complex. The implication is that if you could somehow keep your cells young, you would stay young indefinitely. But this is a hard concept for many to accept, including my friend Hayflick.

I have heard Hayflick speak numerous times about cellular aging and the implications for human aging. He usually begins his lecture by stating that we can't possibly stop the aging process, let alone reverse it. He often uses the analogy of an aging satellite flying through the solar system, gathering damage and "getting older" as the dust and cosmic rays take their toll on the delicate equipment.

"People," he says, "are just like satellites. They get damaged, they get older, and you can't change that fact."

He then goes on to explain his own work, ensuring that the audience understands the mechanisms—and the limitations—of cell senescence and what he calls the "replicometer" our cells contain, which measures cell divisions and enforces cell aging.

Despite his overall skepticism, Hayflick often concludes his talks on an optimistic note, suggesting that there may very well be the potential to ameliorate the ravages of human aging.

Hayflick's replicometer, we now know, is the telomere. And the potential to ameliorate aging exists in an enzyme called telomerase, which affects telomere shortening.

And yes, the research now suggests that if we can alter telomere length, we might be able to slow, possibly even reverse, aging.

Telomeres, Telomerase, and Cellular Aging

Telomeres were first discovered and named by the American geneticist Hermann Muller in 1938, from the Greek words *telos* (end) and *meros* (part). Two years later, cytogeneticist Barbara McClintock

described telomeres' function—protecting the ends of chromosomes in certain cells in multicellular organisms. McClintock later
received a Nobel Prize for her work.

Telomeres comprise the last several thousand pairs of bases
(nucleotides)[5] at the end of each of our chromosomes. The metaphor
often used is that of an aglet, the hard plastic tip on a shoelace.
Each telomere is made of a specific repeated sequence of bases:
TTAGGG (thymine, thymine, adenine, guanine, guanine, gua-

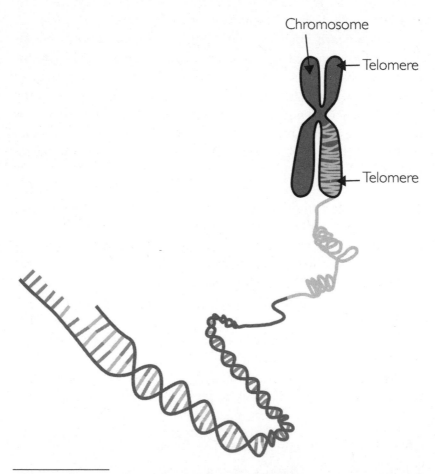

Chromosome

Telomere

Telomere

---

[5] These bases, or nucleotides, are the genetic letters that spell out the genes on
our chromosomes. There are only four DNA "letters"—T, A, G, and C—but these
are sufficient to spell out all of our genes.

nine), which varies only trivially, if at all, between species. Because these sequences don't code for a protein, they are often considered "junk DNA." But this mistakes their critical function. The telomere comprises a tiny part of the total chromosome, but its importance is profound.

Although no one recognized it at the time, the next critical theoretical step toward understanding the role of the telomere was taken by Russian scientist Alexey Olovnikov in 1971. Olovnikov, who lived—and still lives—in a small flat in Moscow, was riding the subway one day when he was struck by the similarities between chromosomes and subway trains. He began to wonder about how chromosomes are copied when cells divide and realized that there was a problem.

A cell uses enzymes known as DNA polymerases to replicate the DNA that makes up the chromosome. But those enzymes have to "hold on" to part of the old chromosome as it begins duplicating the genes, rendering the DNA polymerase incapable of duplicating the part of the chromosome directly "underneath" it—just as a subway car might be used to lay new track, but can't lay track directly under itself.

Imagine someone trying to copy *you* with a portable scanner. The person grabs you by the hand and holds on tight as he waves the scanner from your head to your feet. If he tries to copy your hand, however, he has to let go, and you'll run away. If DNA polymerase "lets go" of the part of the chromosome it has grabbed in order to copy it, the chromosome would simply drift away.

Because the DNA polymerase can only replicate in one direction and must always hold onto that tiny piece of the chromosome, it can never return to copy the nucleotides it missed.

Olovnikov's sudden insight proved entirely correct. While most of the chromosome is copied during replication, a tiny piece is always lost; each time the chromosome is copied, it gets a little shorter. The part of the chromosome the enzyme holds onto, as it turns out, is the telomere. Because the part of the telomere the enzyme is "grasping" can't be copied, it makes the new telomere just a tiny bit shorter than the original. When you are very young—or when your cells are young—the telomere is perhaps 15,000 base pairs long. By the time these cells reach the end of their ability to divide, your telomere might be only 8,000 base pairs long. Olovnikov suggested that telomere shortening was the mechanism of the Hayflick Limit.

At the same time, Olovnikov knew that some cells never age. This includes single-cell organisms, and also germ cells and most cancer cells. There had to be a way for these cells and others like it to "come back later" and duplicate the end of the chromosome that was missed initially. The enzyme that does this—that re-extends the telomere—is called telomerase. It allows certain kinds of cells to reset their telomeres to their full length, so that those cells can continue to repair themselves and to divide indefinitely. Only in cells that don't express telomerase (i.e., most somatic cells) does the telomere shorten with each division.

Telomerase was proven to exist and named in the 1980s, when University of California, Berkeley researchers Elizabeth Blackburn and Carol Greider isolated the enzyme in the protozoan ciliate *Tetrahymena*, an organism that looks like a very small, very delicate jellyfish.[6]

They, along with Harvard Medical School professor Jack Szostak, won the 2009 Nobel Prize in medicine and physiology for their work on telomerase. Olovnikov was not included in the award.

Despite the obvious correlation of telomere length with cell aging, the question of causation remained open until 1999, when re-lengthening of telomeres was shown to reset cell aging in the laboratory.[7] Until then, suggestions that telomeres might play a central role in age-related disease were generally rejected out of hand. This was partly because there was little data proving causation, partly because the link between telomere shortening and cell aging was poorly understood, and partly because it is never easy for people—even scientists—to accept a radical new idea. In the case of aging, we were slowly being forced to reassess everything we knew.

---

[6] Shelton, D. N., et al. "Microarray Analysis of Replicative Senescence." *Current Biology* 9 (1999): 939-45.

[7] Shelton, D. N., et al. "Microarray Analysis of Replicative Senescence." 939-45.

OLOVNIKOV IN MICHIGAN

Born in 1936, Alexey Olovnikov left Russia only once—a brief trip to East Germany—prior to the late 1990s. That is when he flew from Moscow to New York to Michigan, where he joined my wife and me for dinner. I picked up Alexey at the airport, stopping on my way home at a grocery store, where he was astounded at the abundance and variety. Our home is far from extravagant, but we felt a little self-conscious about the relative opulence of our American lifestyle. As I was grilling a steak, a lightning storm blew through the area, leaving us suddenly without power, light, and water. As I fumbled to light a candle and my wife tried to rescue dinner in the dark, Alexey peered across the table at me and said in his thick Russian accent, "You know, Michael, this is not so different from Moscow . . ."

Gene expression determines the way cells use their chromosomes to generate proteins and other key molecules. Young cells have a young pattern of gene expression; old cells have an old pattern of gene expression. Each shortening of the telomere slows the rate of gene expression. As a result, the rate of DNA repair and molecular recycling gradually slows down, increasing damage to DNA and other molecules, such as proteins, lipid membrane molecules, and all of the building blocks that make a young cell work so well. Cells eventually become both dysfunctional and unable to divide any further. They can neither do their specialized jobs nor replace cells lost in the tissue around them. Small wonder that as we age our skin becomes thin and the linings of our joints fail.[8, 9, 10, 11, 12]

---

[8] Hayflick, L. "Intracellular Determinants of Cell Aging." *Mechanisms of Aging Development* 28 (1984): 177-85.

[9] Hayflick, L. "Cell Aging." Chapter 2 in Eisdorfer, C. (Ed.), *Annual Review of Gerontology and Geriatrics, Volume 1*. Springer Publishing, 1980.

[10] West, M. D., et.al. "Altered Expression of Plasminogen and Plasminogen Activator Inhibitor During Cellular Senescence." *Experimental Gerontology* 31 (1996): 175–93.

[11] Shelton, D. N., et al. "Microarray Analysis of Replicative Senescence." 939-45.

[12] Roques, C. N., Boyer, J. C., and Farber, R. A. "Microsatellite Mutation Rates Are Equivalent in Normal and Telomerase-immortalized Human Fibroblasts." *Cancer Research* 61 (2001): 8405-07.

SOMATIC VS. GERM CELLS

All of the tissues of animals and plants are made up of somatic cells, except sex cells, or germ cells. In humans, these are the sperm cells in men and the ova in women. Most somatic cells do not express telomerase, and with every division, their telomeres shorten. Stem cells and cancer cells are an exception, and can express telomerase, allowing them to maintain telomere length despite repeated cell division.

Cells with telomerase can maintain themselves indefinitely. Cells without telomerase slowly go downhill, failing to repair damage, failing to recycle molecules, and failing to divide. Whether they die or simply become quiescent and ineffective, the result is the same: tissue failure and clinical disease.

The Telomere Theory of Aging

Every human begins as a fertilized egg, the union of two germ cells. The fertilized egg divides rapidly, and these new embryonic stem cells differentiate into all the different cell types of the body.

TELOMERASE AND CANCER

Telomerase doesn't cause cancer, but may be necessary for cancer cells to divide. Because cancer cells generate telomerase, they can divide indefinitely, which is partly why they are so dangerous. In 1951, scientists harvested cervical cancer cells from Henrietta Lacks, an African-American woman living in Virginia. These so-called HeLa cells have been used in a vast array of scientific research for decades. Roughly twenty tons of HeLa cells have been grown, demonstrating the agelessness of cancer cells and others that can express telomerase. The story of Henrietta Lacks and the HeLa cells is well told in Rebecca Skloot's book *The Immortal Life of Henrietta Lacks*.

Embryonic stem cells express telomerase and so can divide at will without aging. Babies are born with a few trillion cells, all young and healthy.

Most of these are somatic cells, which soon begin aging with each cell division. A relatively small number of cells, perhaps fewer than one in 100,000, are "adult" stem cells, which can divide at will, creating young cells, but only of a limited type. When a stem cell divides, one of the new cells remains a stem cell; the other becomes a somatic cell. These new young somatic cells have long telomeres, which then shorten with every division, while the stem cell itself resets its telomeres with each division, so it can continue to supply new somatic cells. However, the process usually isn't perfect, so there is a very gradual loss of telomeres even in stem cells. As a result, as we grow older, our stem cells slowly become less able to replace missing somatic cells. While a centenarian's stem cells can still produce new blood cells, for example, they can't produce them quite as well—or as quickly—as when the centenarian was a young adult.

The telomere theory of aging now becomes clear. Most of our cells do not express telomerase, and so their telomeres shorten every time they divide. The shortened telomeres change gene expression—for the worse—and the cells begin to fail. The symptoms of aging we experience—everything from wrinkles to increased cancer risk to Alzheimer's—reflects the aging of these cells. It's that simple and that complex.

What Happens as the Telomeres Shorten

When telomeres shorten, gene expression suffers, and the cell ages. To understand how this works, it's important to understand a bit about cellular function.

Everything in the cell is in flux. Moment-to-moment, molecules in your cells are being produced and destroyed, built up and taken down, recycling continually and ceaselessly. All this destruction and rebuilding may seem wasteful, and in fact the energy cost is enormous. But as a result, most molecules in the cell are fairly new, and therefore likely to be in good shape and functioning perfectly. The cell works very hard to make sure that every molecule is functioning as it should.

While it may seem more efficient to repair damaged molecules rather than replace them, the cell doesn't do this for the same reason you generally don't repair your broken cell phone. If the damage is significant it's actually cheaper and easier to replace your phone— or your molecules.

In many cases, the creative destruction of molecules is biased toward damaged ones, but not absolutely so. While your body can often recognize a damaged molecule and tag it for priority destruction, all molecules undergo recycling, if at varying rates. Perfectly functional molecules are also continually being broken down and replaced by other perfectly normal molecules.

This system of continual recycling is very effective, but there is a downside: It costs a great deal of energy to keep replacing molecules. On the other hand, if the recycling process slows down, the pool of molecules becomes increasingly dominated by damaged ones. As we will see, this problem lies at the heart of aging. Young people have high metabolic rates and are continually renewing their molecules. Old people have slower metabolic rates and don't recycle nearly fast enough.

Consider what would happen if cell-phone contracts were shortened from two years to two months, so that everyone received a new phone every other month. In all likelihood, one would *never* have a damaged phone. In a pool of a thousand cell phones, the odds are that almost all of them would work perfectly well on any given day, since no one cell phone would be more than two months old. But this would be prohibitively expensive.

But what about the opposite? If the cell phone contracts lengthened from two years to two *decades*, eventually most people would have cell phones that didn't work. So we have two extreme options: We can pay a huge amount and ensure that our phones always work, or we can pay very little and find that our phones never work.

In the case of living cells, the same thing happens to our molecules. The speed of molecular replacement determines the level of function of the cell. Young cells replace molecules quickly, and most molecules are functioning perfectly. As telomeres shorten, gene expression changes, resulting in slower replacement of needed molecules. The consequence of slower molecular recycling is medical disaster. If recycling is too slow, most enzymes—the workhorses of our cells—no longer work. Most proteins are defective, most lipids form leaky membranes, and, overall, nothing works well.

This is the aging cell.

The central problem with the aging cell is not that the rate of damage goes up. Nor is it a case of damage simply accumulating because "things wear out." The problem comes when the rate of this recycling slows down and, as a result, the damage gradually accumulates. Cells still work, but they become inefficient and prone to fail, as do cellular products such as intracellular matrix (e.g., skin collagen) or bone, where osteoporosis results. And as the cells and their products fail to perform well, the likelihood of clinical disease climbs steadily until the organism itself fails.

To put it succinctly, cells don't age because they've become damaged; cells become damaged because they've aged.

REPAIRING DNA

There is only one type of molecule that your body ever repairs, and that's your DNA. DNA molecules—the critical and sole source of molecular templates for all other molecules and the blueprints for everything in your body—are continually being checked, repaired, and then rechecked. Damage is never tolerated. The process of monitoring and repairing your DNA is complex and metabolically costly, but it's essential. When the cell identifies damage, it either repairs the problem or else stops all future cell divisions to prevent the error from being passed on to daughter cells. Occasionally, this safety mechanism fails, and damaged DNA is passed on to daughter cells. All too often these are cancer cells. The repair of DNA therefore has a high priority, as the cost of not repairing the DNA is often the death of the entire organism.

Relationship to Other Theories of Aging

Look at what this means. Cells do *not* accumulate damage passively. Rather, as cells get older, the rate at which they repair and replace damage slows down. With this understanding, the relationship between the telomere theory of aging and the various theories discussed in Chapter One is clearer.

In the wear and tear theory, cells age because they passively accumulate damage. But in fact, wear and tear occurs in all cells, regardless of cell age. It only causes problems when the damage is not repaired quickly enough. Young cells keep up with the damage, old cells fall behind.

The telomere theory explains why some cells are able to stay young and avoid wear and tear. We now know that young, healthy cells can stay ahead of wear and tear and—if they express telomerase—do so indefinitely.

The mitochondria free-radical theory also has much truth to it. As noted earlier, energy production creates free radicals that can damage molecules, including DNA. Because it's imperative for cells to keep their DNA intact, having the DNA confined and protected within the nucleus and the free radicals trapped within the mitochondria drastically decreases the rate of genetic damage.

The free-radical theory argues that young mitochondria are efficient, with a low rate of free-radical production, but with aging, the mitochondria become inefficient and produce more and more free radicals, which damage the cells, ultimately causing the entire organism to age. This explanation is odd in that it ascribes aging of the organism to aging of the mitochondria, which begs the question of why mitochondria age in the first place. But even more problematic, the very same mitochondria function for millennia in germ cells without problems, because germ cells can express telomerase, maintain young mitochondria, and therefore stay young and healthy. Just like cells themselves, some mitochondria get old, while others can remain young indefinitely.

Telomere theory explains these issues. Human mitochondria have their own separate set of thirty-seven genes in a ring chromosome—no ends, no telomeres. Why, then, do some mitochondria appear to age? Most of the proteins needed for mitochondrial function are actually encoded not by genes in the mitochondria, but by genes in the cell's nucleus. The proteins are then imported into the mitochondria. So mitochondrial function is actually dependent on the nuclear chromosomes, whose telomeres shorten and whose pattern of gene expression changes over time. As the cell ages, its ability to supply the mitochondria with all of the proteins it needs decreases, and as the mitochondria fail, the production of free radicals goes up. Aging also means slower replacement of the lipids that make up the mitochondrial and nuclear membranes, allowing free radicals to make their way out of the mitochondria and to the cell's DNA more easily. Moreover, the scavenger molecules that capture and destroy free radicals also begin to fail with age. As we age, we produce more free radicals, they escape more easily, we aren't as good at capturing them, and we aren't as good at repairing the damage once they do. And all these problems—the aging damage caused by free radicals—can be traced back to the telomere.

The fundamental problem is not aging mitochondria. It's a senescent pattern of gene expression driven by shortening telomeres that allows free radicals to unravel our cells.

The telomere theory of aging can now be put in one sentence: *Cells divide, telomeres shorten, gene expression changes, cellular repair and recycling slow down, errors slowly accumulate, and cells fail.*

Misconceptions About the Telomere Theory of Aging

Living cells were discovered three and a half centuries ago by Robert Hooke, a British "natural philosopher," who called them cells because, when he first viewed plant cells in his microscope, he felt that they looked like the cells in a monastery. It was Hooke who first showed that large life forms—such as humans—were not single, confluent organisms, but were made up of innumerable tiny cells.

This observation was a turning point in both biology and medicine. Prior to Hooke, the body was seen as a single living gestalt, or a collection of different organs and tissues all of which shared some mysterious life force, the *élan vital*. The concept of cells, however, led to an entirely new view of how life functioned, and it laid the groundwork for modern medicine.

Over the next few centuries, as the microscope enabled people to look directly at cells, the central tenet of biology, vitalism, was slowly replaced by cell theory. Biology became focused on a single, fundamental building block: the cell. In the twenty-first century, cell theory sounds self-evident, but in some odd ways, the tendency to think in terms of vitalism remains in our approach to both theory and clinical intervention.

The best example of this is the way we view aging. A hallmark of medical pathology is that all diseases are ultimately *cellular* diseases. Once we understand the pathology within a cell and how it causes problems for neighboring cells, then we have a basic understanding of the disease. Yet many people still cling to the notion that aging is not something that happens *within* cells, but something enigmatic and gestalt-like that happens *between* cells, while the cells themselves are merely innocent bystanders.

Disease begins primarily within cells and results in secondary problems between cells, not the other way around.

Overcoming this misconception is critical to general adoption of the telomere theory of aging. But it's only one of the misconceptions that needs to be addressed. No theory I can think of has been plagued by more misconception and confusion. Let's consider the major ones.

Misconception #1: Telomere Length Defines Aging

The most common misconception about the telomere theory is that telomere length defines or causes aging. The truth is, an organism's telomere length has almost nothing to do with how long it lives or how fast it ages. As many researchers point out, some animals, such as mice, have long telomeres but short lifetimes, while other animals, such as humans, have much shorter telomeres but longer lifetimes.

Telomere theory doesn't suggest that telomere length controls aging: Telomere length is irrelevant to aging. Rather, *changes* in telomere length control cell aging. The data consistently support this observation. The key question isn't how long your telomeres were at birth, but how much your telomeres have shortened. It's the shortening that alters gene expression.

Observations of the change in telomere lengths from birth to senility in mice and other organisms show clearly that telomere shortening—or rather, the way in which shortening telomeres cause changes in gene expression—is the driving force in the aging of the whole organism.

This is part of the reason why measurement of telomere length has limited predictive clinical value. Only if you know the average telomere lengths for a particular type of cell in a particular species can a single telomere length be used to help assess body function and pathology. For example, if I know that teenaged humans have an average telomere length of 8.5 kbp[13] in their circulating white cells but that this usually falls to 7.0 kbp by age eighty, then finding that your white-cell telomeres have a length of 6 kbp tells me you're in trouble. The length, 6 kbp, is irrelevant unless we know the context. It's not the length, it's the *change* in length.

Also, the validity of telomere length depends on the type of cell chosen. Some cells show telomere shortening as we age, some don't. Many cells—such as those lining the arteries, the glial cells in the brain, or cells found in the blood, skin, GI endothelium, and liver—divide throughout their lifespan. But many others—such as muscle and nerve cells—generally cease dividing before birth and hence have relatively stable telomere lengths as we age. We might expect to find clinical value in learning how the telomere lengths have decreased in cells lining your coronary arteries, but it would be almost useless to measure the telomere lengths in your cardiac muscle cells. Likewise, it would be useful to follow telomere shortening in your microglial cells, but not in following telomere shortening in the brain cells that they support.[14]

---

[13] Kilo-base pair, a unit of measurement of DNA or RNA length used in genetics, equal to 1,000 nucleotides.
[14] Actually, even in adults, some neurons and some muscle cells divide, but very rarely.

Misconception #2: Cells Die Because Telomeres Unravel

Despite what you might have seen on television health programs, telomeres do not unravel. This common misconception derives from the common analogy of telomeres as aglets, the plastic end caps on shoestrings. The implication of this metaphor is that when you get older, the plastic aglet of the telomere wears away and all the strands that make up the DNA unravel, causing your chromosomes to come apart, killing the aging cell.

But this isn't what happens.

(Note: Originally, my publisher had a shoelace on the cover of this book, because the shoelace's aglet is a powerful metaphor for the telomere. But the metaphor was wrong, so the cover changed as shown.)

In fact, chromosomes never unravel, because deterioration never gets that far. Cellular dysfunction reaches a tipping point long before the telomere is used up. Only in the most extreme cases, such as the fifth generation of "telomere knockout" mice (which cannot express telomerase), do cells ever lose all their telomeres. It simply doesn't happen in normal aging.

In reality, your chromosomes actually remain in pretty good shape, even if you live to be 120. The only time they actually fray is during decomposition.

Likewise, the idea that telomere shortening is what kills the cell is usually inaccurate. Cells with short telomeres certainly don't work very well, but that doesn't mean they're dead.

Misconception #3: Aging Diseases Can't Be Related to Telomeres

Almost invariably, someone will argue that telomeres couldn't possibly cause heart disease or Alzheimer's dementia. Usually, this argument comes from a perfectly rational academic scientist whose grasp of biology is magisterial, but whose grasp of clinical pathology is much less so.

In the case of heart disease, they point out that heart muscle cells, cardiomyocytes, almost never divide, and so heart disease can't possibly result from telomere shortening.

But the pathology is more complex. Saying that telomere loss can't cause heart attacks because heart muscle cells don't lose telomeres is like saying cholesterol can't cause heart attacks because heart muscle cells don't accumulate cholesterol.

It's not changes to the cardiomyocytes that lead to heart disease, but changes in the coronary arteries—the vascular endothelial cells—which lose telomeres *and* accumulate cholesterol. The underlying pathology lies in the arteries, not in the muscle. The fact that cardiomyocytes don't divide is irrelevant to the pathology of heart disease.

The same criticism—with a similar misunderstanding of pathology—is used in regard to Alzheimer's dementia: Neurons almost never divide, so Alzheimer's dementia can't possibly be due to telomere shortening.

While it is roughly accurate to say that adult neurons don't divide, the microglial cells that surround and support those neurons divide continually, and their telomeres certainly *do* shorten with age. Microglial telomere shortening correlates with Alzheimer's disease and appears to precede the onset of several hallmarks of dementia, including beta-amyloid deposition and the formation of tau protein tangles.

It's useful to make a rough distinction here between direct age-related pathology and indirect age-related pathology. Alzheimer's and heart disease are examples of indirect pathology, where neurons and the cardiomyocytes are "innocent bystanders." Direct aging means that aging cells cause pathology in their own tissue; indirect aging means that aging cells cause pathology in a different tissue, or different cell type. This distinction will prove especially useful as we move into subsequent chapters and all the more so as we enter the realm of clinical intervention using telomere extension. In Chapter Five, I discuss telomere shortening and direct age-related pathologies; in Chapter Six I discuss telomere shortening and indirect age-related pathologies.

Moving from Theory to Intervention

All truth passes through three stages. First, it is ridiculed.
Second, it is violently opposed. Third, it is accepted as being
self-evident.

— Arthur Schopenhauer

Human biology is immensely complex, and with regard to aging the question of causation even more so. Causation operates at many levels, and, as we have seen, the nature of aging can be explained in many ways.

In one sense, aging isn't caused by anything. It's a dynamic and complex cascade of events with no single beginning and no event we can point to as *the* cause. We might legitimately say that aging is caused by free radicals or accumulating damage to the organism or any number of other accurate but misleading causes. We could try to pin causation down by saying that aging is caused by changes within cells as a result of inadequate repair and recycling resulting in changes in the organism and a growing probability of certain typical diseases.

But it might be most accurate to say that telomere shortening doesn't so much cause aging as gradually expose our underlying genetic weaknesses or our predispositions to disease. Aging doesn't cause disease, but it does increase the chances that your familial risk of heart disease, for example, will surface and cause pathology or mortality. Aging doesn't cause the heart attack, but it does make it increasingly more likely. In a sense, we might view telomere shortening—or aging—as sailing on a lake in which the water level gradually drops, slowly bringing rocks and shoals closer to the surface. The closer to the surface these shoals of genetic risk become, the more likely you are to find yourself shipwrecked on one of them. You might be lucky enough not to be at risk for atherosclerosis, but there will still be some other shoal for you to run aground on. Sooner or later, as your telomeres shorten and aging continues, some unseen risk will eventually arise, resulting in disease and finally death.

But these discussions of causation are not my main focus. As a doctor, my concern isn't causation, but intervention. What I am interested in is the practical, defined, and testable.

The critical question is this: What is the *single most effective point of intervention* to cure age-related disease? That's the practical approach, both for me as a clinician and for all of us who live long enough to age and get diseases. Also, it defines the issue well in that I don't actually care whether the *cause* of aging is free radicals, cosmic rays, methylation, telomere shortening, or anything else. What I want to know isn't the cause but the cure. Which of these factors, if any, can be used to intervene in aging and which of them will do best help us to cure or prevent actual disease? This approach is inherently testable. If we suggest that telomeres are the most effective point of intervention, then we can test that notion by supplying cells with telomerase, for example.

The telomere theory of aging posits that the key to intervening in age-related diseases is employing telomerase to re-lengthen telomeres, thereby restoring gene expression to its most healthy state. Until now, most interventions have either focused on symptoms, such as pain, or on individual problems that result from changes in gene expression, such as inflammation. This narrow focus results in many clinical failures that attempt to cure age-related diseases. In the case of Alzheimer's dementia research, for example, the majority of clinical trials have aimed at beta-amyloid proteins and related molecules, as well as tau proteins. The failure of these trials is to be expected, because they don't deal with the broader issues that drive aging cells to fail in the first place. These trials aim at effects (such as beta-amyloid plaques), rather than causes (such as microglial cell aging). It's like trying to treat a fatal viral infection by treating the fever; we may lower the temperature, but the patient still dies of the infection. When we want to cure a disease like Alzheimer's, it's not enough to aim at beta-amyloid deposits. If we want to cure Alzheimer's, we need to reset gene expression in the microglial cells that would normally prevent beta-amyloid deposits in the first place. Small wonder that none of the past efforts has succeeded. We're aiming at the wrong clinical target.

Moving Toward Consensus on Aging

Although telomere theory is consistent with the actual data and able to explain the gamut of biological aging and age-related pathology, it is often misunderstood, even by researchers. Many have criticized the approach without understanding the theory and—in some cases—without understanding clinical pathology. This is understandable, as the theory is actually easy to misconstrue and many experts in basic biology are, as we saw in Misconception #3 above, not experts in human disease. In addition, there has been a dearth of published sources that explain the theory as a whole.

But the tide is turning. I find that older academics are critical of the theory, while the younger academics increasingly accept it as a given, even if neither has an accurate theoretical understanding.

For my part, these academic discussions of causation are moot exercises in philosophy, enjoyable for conversations over wine but not particularly productive. In recent decades, I've focused more on the science leading directly to clinical intervention. I'll discuss this in detail in Chapter Four.

But first, I want to take an interesting detour. Most of this book describes *how* we age, but in the next chapter I ask what might be a surprising question: *Why* do we age?

CHAPTER THREE

Why We Age

As you read in the last chapter, we age because every time our cells reproduce, our telomeres shorten. Each shortening of the telomere leads to increasingly poor cellular functioning. This, in a nutshell, is the telomere theory of aging, and it tells us an enormous amount about how we age, as you will see in the rest of this book.

In this chapter, however, I want to take a brief diversion to ask another question: Why do we age? Earlier theories of aging tend to avoid this question. If we are aging because of wear and tear or free radicals, the answer is clear: We age because aging is inevitable. Despite our bodies' best efforts, the accumulated damage eventually overcomes our ability to repair that damage.

But the telomere theory of aging makes the question of why we age far more interesting. Our cells all have the gene for telomerase. They could express telomerase, just as our germ and stem cells do. But they don't. Apparently, our bodies, rather than aging as the result of an inevitable physical process, are, in fact, *designed* to age. Our bodies age on purpose.

Evolutionary Thinking

Evolution is cleverer than you are.

— Leslie Orgel, evolutionary biologist

Any time we ask why something happens in biology, we are asking an evolutionary question. The functioning of every living thing on this planet is the result of billions of years of evolution, and virtually every aspect of your body is the result of the relentless workings of evolution. If aging didn't make good evolutionary sense for a species, then organisms wouldn't age. Somehow, aging makes it more likely that our genes will replicate and our species will survive.

Asking why can feel like an endless children's game, but in biology the whys generally end with an evolutionary explanation. For example:

Q: Why do we get hungry?

A: Because we haven't eaten in a while.

Q: Why does not eating make you hungry?

A: Because when you haven't eaten your body produces less leptin, and that causes you to feel hungry.

Q: Why does your body do that?

A: Because animals that didn't get hungry didn't work as hard to find food, and didn't survive to reproduce, while animals that got hungry survived. People inherited this trait from their animal ancestors.

Note that it's very tempting to say that your body "wants" you to eat or that evolution "wants" you to eat. I'll use this shorthand occasionally, but it's important to remember that saying "want" is only convenient shorthand. Evolution doesn't "want" anything, but if you don't eat, then your genes don't survive. In this chapter we are asking, "Why does our body want to age?" but this is shorthand. The real question is this: Why did animals that age out-survive and out-reproduce animals that didn't age? Why does aging make a species (and its genes) more likely to survive?

Because multicellular animals began aging billions of years ago, any answer to the question of why we age is inherently speculative. But asking the question can teach us a lot about evolutionary reasoning and the nature of the evolutionary process.

Evolutionary Cost-Benefit

Why aren't lions faster? Lions can run at an impressive fifty miles per hour for short bursts, but their prey are equally fast. Wildebeest can run at fifty miles per hour as well. Zebras and Cape buffalo can run at around forty miles per hour. Because lions survive by chasing down prey, why haven't they evolved to run as quickly as a cheetah, which can run seventy miles per hour?

Because, as with everything in evolution, there are tradeoffs. Cheetahs can reach incredible speeds because their slender bodies, small heads, and long, thin legs are very aerodynamic. They have oversized hearts and large lungs and nostrils to allow their muscles to stay oxygenated at these speeds. But these benefits come at a price. The cheetah's small head means smaller teeth and weaker jaws than most predators.

A faster running lion might find it easier to catch prey, but may need to be less muscular and more aerodynamic. This faster lion might fall victim to competitions with other lions and never live to reproduce.

The point is that to understand why evolution takes a certain direction, it's necessary to look at tradeoffs between the costs and the benefits.

This brings us to the question of aging. Wouldn't it be a big evolutionary advantage to live and reproduce indefinitely? Animals that didn't age could theoretically produce many times more offspring than their aging competitors.

It turns out that aging has a much lower cost than it might seem. While there is great variation in lifespan from species to species, most animals do not live a full lifespan and die of old age. Starvation, interspecies competition, predation, disease, and cancer kill most animals long before they reach old age. Aging isn't a factor in their deaths. The "cost of aging" only applies to those animals who reach a point where aging is a factor in their death.

There's another factor that is a little more subtle. Organisms function in their own ecological niches, which can only support so many of each species. The limiting factor in the number of deer, for example, isn't the rate at which deer can reproduce. It's based on the availability of food and prevalence of predators. If a given area can support a population of 1,000 deer, what would happen if another 1,000 deer were suddenly added? Starvation and predation would soon reduce the population back to 1,000.

With this in mind, imagine a small group of ageless deer that reproduce indefinitely. These deer are a small subset of the total population of deer, which age normally. The aging deer, like all species, continually evolve in response to changes in the environment. But the ageless deer are producing offspring that represent an earlier evolutionary phase. With each generation the offspring of the ageless deer are less fit than the offspring of aging deer. These ageless deer would be quickly be crowded out.

This last point is the key issue in the evolution of aging. If a species has a long lifespan, then it can't adapt as quickly as a species with a shorter lifespan. It's a bit like the turning radius of a car—if the turning radius is shorter, the car can turn sharper corners. If a species lives a long life and has offspring late in life, then the "turning radius" of the species may not keep up with rapid changes in its physical or biological environment. If the temperature, oxygen, pH, or some other physical aspect of the environment changes, a species needs to change with it. If the biological competition or a prey species changes quickly, then once again, a species with a shorter lifespan can adapt more quickly and is more likely to survive. On the other hand, if the environment—physical or biological—is stable, then longer lifespans are advantageous to survival. Lifespan and the rate of aging have to be finely tuned not just to the environment, but also to the rate at which the environment changes.

So the benefits of "agelessness" are much lower than it might seem, for two reasons: Most deaths occur before aging decreases fitness, and agelessness slows the rate of evolution. Aging has benefits to a species, but the costs to an individual—aging and disease—are severe.

Historically, we've sometimes assumed that aging was simply part of being multicellular. As it turns out, some multicellular organisms (like hydra) don't age, while some unicellular organisms (like yeast) do age.

The Multicellular Dilemma

Multicellular life first evolved around one billion years ago, after an estimated 2.6 billion years during which only single-celled organisms lived. Early multicellular life was in the form of cooperative colonies, in which single-celled organisms could thrive better than they could on their own over time.

Multicellular life eventually learned to differentiate cells to allow for more sophisticated organisms with specialized germ cells for reproduction. Consider how radical this change was for the cells of multicellular creatures. For billions of years cells had evolved to survive and reproduce. The single-celled organisms that reproduced most rapidly and most successfully crowded out those single-celled organisms that were not as aggressive.

Now, as part of a multicellular life form, cells had to learn a very different way of behaving. Cells had to operate responsibly to perform their roles in support of the whole organism. They had to divide only when needed for the benefit of the organism. A cell that divides too rapidly—when the organism doesn't need it to divide—is a cancer cell, which kills the organism. Organisms with cells that reproduced willy-nilly were selected against; organisms that carefully controlled their cells survived and prospered.

Multicellular creatures evolved to control their cells' reproduction. What was the mechanism of that control? Part of that mechanism was cell aging.

The Hayflick Limit provides a harsh but powerful tool for controlling cell reproduction. After a certain number of divisions, cells simply couldn't reproduce any further. With each cell division, the telomeres shorten, and after forty[1] or so divisions, most cells simply can't divide any further. This mechanism of control came at a price—aging and death from aging—but, as we saw earlier, this price wasn't all that high, evolutionarily speaking, and may have had some benefit in fine-tuning the rate at which a species can adapt to environmental changes.

Why Do We Age?

While the complete answer to why we age may never be known, it seems quite likely that aging was a product of evolution—a tool to enhance a species' ability to adapt quickly to environmental change. So if evolution "chose" aging, can scientists develop tools to "unchoose" it in whole or in part? This brings us to the next chapter, in which we leave theory behind and examine the progress we've made applying the telomere theory of aging to improving health and lengthening lives.

---

[1] Actually, forty divisions are typical for human fibroblasts, but each species and cell type has a specific replication limit.

CHAPTER FOUR

The Search for Immortality

Old theories never die; their proponents do.

— Anonymous

For twenty-five years, medicine has teetered on the brink of transformation. By 1990—thanks to Hayflick, Olovnikov, Harley, and others—we had a basic understanding of cell aging but were only beginning to suspect that cell aging explained *human* aging. The idea that telomeres might help us cure age-related disease was at best a cautious dream.

New theories require data, but also demand determination and patience. The telomere theory of aging was no exception. As with any new theory, it wasn't enough to be *right* or to have supportive data. It takes time for a new theory to gain acceptance. As we approached the turn of the century, a new generation of scientists and physicians began to work and build their careers with a new view of how aging and age-related diseases work.

One: Promising Beginnings

Geron, a biotech company based on telomere research, was founded by Michael West in 1990. Mike West saw the implications of cell aging for human disease. Geron's initial goal was to find a way to intervene in the aging process. Mike showed genius in his grasp of the aging process, and he conveyed his vision to his investors in a valiant effort to move the burgeoning field of telomere biology from the laboratory to the clinic.

Geron Corporation—the name comes from the Greek root word for aging—was the first biotech corporation aimed directly at preventing and reversing human aging. Geron hired Cal Harley as their chief scientific officer in 1992. Eventually, the company held most of the major patents for potential clinical interventions that arose from telomere research. Geron had three major thrusts:

- Using telomerase *activation* to treat aging and age-related diseases
- Using telomerase *inhibition* to treat cancer
- Developing *stem cell* therapies

The ideas that aging was driven by telomere shortening *and* that we might be able to change the aging process were novel and, unsurprisingly, difficult for many people to accept. Even for those who saw the potential, there was little solid supporting data during those first ten years. Investors and board members could easily grasp the potential for a cancer therapy, but they had a harder time believing in the idea that we might reverse the cellular aging process and create profitable cures for age-related diseases. Even some of those who were key players in the field, Len Hayflick among them, remained leery of the idea that telomerase might play a role in human aging. This inherent conservatism—and the cautious fiscal decisions that it drove—reflected the views of most of the scientific community throughout the 1990s.

Nonetheless, the first ten years at Geron were heady times for those of us who saw the potential of telomerase for treating human aging and disease. During that time, Geron focused on understanding how telomerase works, identifying its co-factors, amassing data on the correlation between telomere length and cell aging in multiple cells and species, and other side issues. They conducted important research on telomeres in cancer cells and stem cells. And in 1999 Geron finally clinched the case, showing that telomere shortening doesn't merely correlate with cell aging, but causes it.

It was always hard to keep up with Mike West, and not just intellectually. He might be in his office finishing a patent, he might be in the laboratory counting cells, but he might just as likely be talking about stem cells in Paris or discussing aging cells in Singapore. A world map in Geron's hallway was covered in multicolored pins, trying to keep track of him. Where in the world was Mike West?

Wherever Mike was, he was always a step ahead of everyone else.

In 1993, Geron invited me to fly out to California, gave me access to their unpublished data, much of it proprietary, and had me give several talks about the clinical potential of telomerase for aging and disease. Even then, some of us clearly saw the enormous potential for curing diseases. Seeing the importance of what was happening at Geron, the CEO, Ron Eastman, wanted me to document that history, but the ideas themselves were more important than the company. The book I wrote, *Reversing Human Aging*, was the first history of our growing understanding of why and how aging occurs.

As the only physician working in parallel with Geron, I provided clinical guidance, but I was also one of the few people at Geron who, along with Mike West, was completely committed to the belief that we really could cure age-related diseases. I was the optimist, the theoretician, and the one who told the story to the public. I was the one who saw the broader picture, who could understand the entire theory of aging and its clinical implications. My background—as a medical professor, a university teacher of the biology of aging, and a researcher—enabled me to clearly explain the general telomere theory of aging to the clinical community and to the public.

On my trips to Geron, I stayed with my friend Cal Harley, their chief scientific officer, and his wife at their lovely home in Palo Alto. We spent considerable time discussing the science and my clinical understanding of the aging process. Over dinner one evening, I pointed out that our understanding of telomeres might allow us to construct a single, unified theory of aging. I argued that telomeres were the most effective point of clinical intervention for the diseases of aging, allowing us to prevent and cure most age-related diseases, for which medicine had little or nothing to offer. I was surprised to find that Cal generally disagreed that telomerase could become an effective and innovative medical therapy. I asked him to put a number on how important telomerase might be in driving the aging process. His answer: "No more than fifteen percent." Despite being the key scientific leader at Geron, Cal was still pessimistic about the clinical potential of telomerase, as were many of his Geron colleagues. Ironically, while Geron's vision gradually became more limited over the next decade, Cal himself became more optimistic.

These were intoxicating times.

Most of us had the conviction that something remarkable was in the wind. For the scientists at Geron, there was the knowledge that telomeres might control aging in cells. For me and a few others, there was a strong suspicion that the implications went far beyond aging cells. As a professor of medicine, I had spent years not merely trying to understand human aging, but also trying to find a way to cure the diseases of aging. Telomere biology and cell aging suggested a way to fit everything we knew about the aging process into a single clear, consistent picture, one that an increasing number of us began to share.

There was something far more important at stake than a scientific understanding of aging. Just as the physicists of the 1800s thought that they "almost" completely understood the universe— until relativity and quantum mechanics changed everything in the early twentieth century—biologists and physicians had thought that they "almost" understood aging. And now telomere biology was changing everything. But the revolutionary change struck much deeper than just gaining a more sophisticated understanding of cell aging.

There was a paradigm deeply embedded in medicine and biology that no matter how well we might understand aging, it could never really be changed. We might someday fully understand the molecular details of aging. We might even use this knowledge to slow the aging process and provide palliative care for Alzheimer's disease and atherosclerosis. But we could never reverse or even stop the aging process.

Or could we?

That was the astonishing implication of our new understanding of aging.

Not only did telomeres and cell aging explain how we aged, but for those who looked more closely, the implication was that we might reset the telomeres, reverse cell aging, and thereby cure age-related human diseases. Which is to say that we might reverse human aging.

As a clinician, my concerns have never been purely academic. My goals were practical and clinical. I wanted to improve lives. I wanted to find the single most effective point of intervention for age-related disease. I had already been writing a medical textbook about the aging process, but I was so struck by the implications of telomere research and by its absolutely unprecedented implications for clinical medicine that I shelved the textbook. Instead, I decided to write the first ever book to explain how we might reverse human aging.

Geron had given me free access to their data, but in 1995, much of that data was still preliminary. We were sure that telomere shortening played a role in cell aging, but did it cause cell aging or was it a result? It was clear that young cells had long telomeres, while old cells had shorter telomeres, but which came first, aging or telomere shortening? We suspected that telomere shortening somehow changed gene expression and that drove cell aging, but it had yet to be proved.

The telomere theory of aging was elegant and consistent, and the potential for human medicine was stunning and unprecedented, but the data was not yet conclusive. While I could certainly explain the theory and the research data then available, much of the book would have to be speculative rather than factual. Clearly, the book was not simply a scientific textbook—such as I ended up writing a decade later for Oxford University Press—but a book meant for the general public. So I went on to explain not only the limited theory of how cells age, but also a general theory of human aging and what it could mean to all of us.

It was a difficult task.

Aging science is a field in which everyone divides into two distinct camps, each of which distrusts the other. In one camp are the enthusiasts—excited laymen and a few professionals, regarded by most as fringe elements, who go far beyond the scientific data, often claiming that this miracle food or that fashionable drug can prevent aging. The other camp is made up of thoughtful, serious researchers and clinicians who are so leery of being labeled fringe elements that they take great pains to dissociate themselves and their work from the enthusiast group.

The enthusiast groups usually have large annual meetings with dozens, even hundreds, of sales booths and passionate speakers who make grandiose claims with little factual support. These passionate supporters put great effort into trying to change national regulatory restrictions and research priorities to match their own views. The head of the National Institute of Aging once confessed to me that a few of the key individuals in one of these groups "had done more to undermine congressional support for aging research than anyone in this country." He noted that some congressmen had come to assume that the zealots represented the research community as a whole and saw little reason to provide support for what Congress viewed as "nutty ideas." Given this perception, the scientific and medical communities were often afraid of being tarred with the same brush and carefully distanced themselves from these evangelists of the aging community.

Mike West and I once attended a well-known annual conference on "anti-aging medicine." Various presentations claimed that magnets, crystals, meditation, deionized water, or vitamin supplements would reverse the aging process. This wasn't science. It was magical thinking.

Unfortunately, between these two disparate camps there was very little middle ground for consensus, compromise, or considered understanding. However careful I might be in explaining telomere theory and the prospects for altering human aging, I ran the risk of becoming a pariah of one group or the other. The challenge was formidable: How could I explain to the public the research data, the broader understanding of biological aging, and its unprecedented implications for both medicine and society? Most of the public, like most researchers, are committed to assumptions about aging, and the foremost assumption is that aging can never really be reversed.

It was hard to get the message across accurately.

National medical journalists interviewed me about mitochondria and ignored telomeres altogether. Major bookstores thought *Reversing Human Aging* was a how-to book and put it in the Diet section, even though diet has nothing to do with cell aging or telomeres. The media thought aging was related to free radicals. The public thought aging might be slowed down with the right diet. Neither believed that anything could ever really change aging.

Giving a talk at the National Institutes of Health in April of 1996, I found myself facing an auditorium filled with several hundred physicians and researchers who had their own preconceptions about aging.

"When I finish this lecture, anyone who leaves this room thinking that we can reverse human aging is irrational. However, anyone who leaves this room thinking that we *can't* reverse human aging is also irrational. If you're are at all reasonable, you will leave here thinking that you don't know if we can reverse aging or not, but you'd like to see the data. Now let me show you what we know."

Reversing Human Aging was the first book to explain how we might reverse aging, but it also speculated on the unprecedented medical and social implications. If we re-extended telomeres, would we ever age? If we reversed aging, what would happen to the world population? What about the cost of living? What about the ethics of extending lifespan? The book was based on clear scientific data, but it also predicted what we might do clinically within the next few decades. In 1996, I predicted that the first agent capable of altering human aging would be in clinical testing within one to two decades. That became reality only eleven years later. The first human trials of a telomerase activator began in the spring of 2007.

AN EXTRA DIGIT

Geron's chief scientific officer Cal Harley gradually began to believe in the telomere theory's potential for human medicine. At one point, he gave me a gift. He bought a kitchen stool, carefully carved a DNA helix and the letters TTAGGG—the telomere sequence—on the top of the stool, then hand-painted his initials and a set of numerals on the underside: 01994. "Do you know why I added the extra zero in front of the year?" he asked. "Because if you're right about extending the human lifespan, we might need another digit to keep track of the date."

The book's publication was a curse and a blessing. Many readers fell into those two opposing camps that I was already well aware of: enthusiastic believers who read too much into the book, or disbelieving conservatives who refused to take it seriously. Some saw the speculation as facts; other saw the facts as speculation. Nonetheless, a good many people read the book, and as I went around the world lecturing, my goal was to ensure that people understood the theory and its medical potential.

People sometimes read beyond the facts, then become irate at their own exaggerations. While I was lecturing at the Smithsonian Institution, one member of my audience looked skeptical and even annoyed as I presented the research. He was angry as he stood to ask his question.

"*Doctor* Fossel." He scowled up at me. "Isn't it true that none of this work, not one single experiment, has been done except in cells and then only in the laboratory? This has never been tried in a human being! Is that not true?" He stood triumphantly, sure of his ground, knowing he had me cornered.

"Yes, thank you. Well put. I couldn't have said it better or more succinctly." I looked at the rest of my audience. "Next question, please?"

He was deeply disappointed, but facts speak for themselves.

The scientific response to the book was predictable, given the cultural divide within the aging community. Some discounted it altogether; many simply never read it. (Many researchers still don't know who first published the telomere theory of aging.) Some attacked it not for being wrong, but merely for being speculative.

Oddly enough, *science depends on speculation.*

Any good hypothesis needs to be tested, but without a hypothesis you don't know what you are trying to test. I had been quite clear about which data we had and which data we still lacked—and equally clear about what was hypothetical. Theories must be testable or they aren't science. The telomere theory was completely testable, but oddly enough, many of the responses were critical not because my theory was irrational or wrong, but because it had yet to be proved. Nobel laureate Carol Greider wrote this in a letter to me in 1996:

> Telomere shortening is correlated with cell age; there is yet no evidence that telomeres in any way cause cellular senescence. Further, there is no direct link between cell aging and aging of humans.

Some criticism went far beyond this. Two well-known telomere researchers—using identical wording in their two letters to the publisher—stated that my telomere theory of aging was "an insult to science and science education" and demanded that the book be withdrawn—or actually banned.

I was stunned. Maybe the mark of a paradigm-shattering theory is not just its supporting data, but that someone demands that it be banned *before* it can be tested.

The greater the ignorance, the greater the dogmatism.

— Sir William Osler, "the father of modern medicine"

By the mid 1990s, it had become increasingly clear that telomeres were central to cell aging, but the critical issue was what cell aging had to do with human disease. Even more important, could we use this knowledge to cure these diseases? To a cell biologist, aging was an academic question, but outside the laboratory these questions involve actual people with real medical problems. Patients aren't seeking to know how telomerase extends telomeres; they want to know if telomerase extends lives.

Scientists deal with cells, but physicians deal with suffering.

Aging is an abstract concept, but to physicians—and to all of us sooner or later—aging is also dementia, heart disease, painful joints, and fear. The important issue is not cells, not aging, and not even medical diseases, but *intervention*. Can we find a way to cure the diseases that undercut our lives as we age? Can we improve the lives of those we love and care for?

I don't want to *understand* aging, I want to *treat* aging.

I wrote several influential medical articles, the first articles ever to explain the general telomere theory of aging and to suggest that we might cure age-related diseases by using telomerase. I saw telomerase as a way to do this by directly resetting the aging process at the cellular level. My articles in the *Journal of the American Medical Association* (*JAMA*) summarized the data and gave the case for how telomerase could transform our ability to treat age-related diseases. They were the first published, peer-reviewed articles to argue that shortening telomeres cause not only cell aging but also human aging, *and* that telomere elongation could cure age-related diseases *and* actually reverse aging.

The medical community responded favorably, as did many in the academic scientific community. Nevertheless, we needed data.

Within a few years of my pioneering medical articles, Geron proved that telomerase indeed resets aging in cells. Until then, the work had been largely theoretical, and all of the work—both Geron's initial success and the telomere theory of aging—depended on laboratory results. Cal Harley led his team in getting those results. Cal's work emphasized the striking correlation between telomeres and cell aging, but he went much further than that. It was Harley and his colleagues who proved that telomere shortening was not only correlated with cell aging, but that it actually *caused* cell aging.

By 1999, what my *JAMA* articles had proposed had now been proved in Geron's laboratories.

Critically, Harley showed that when telomerase was used to reset telomere length in an old cell to a length typical in a young cell, the old cell became indistinguishable from a young one. In short, changes in telomere length did not simply correlate with cell aging, they were responsible for cell aging, and they could be reset. For the first time in human history, we had actually reversed aging in human cells. Using telomerase, we could turn back the clock, making an old cell young.

In 1999, Geron's scientists published a history-making paper[1] in which they showed that when you reset telomere lengths in old human cells, you reset not only their Hayflick Limit, but also the pattern of gene expression as well. Old human cells looked and acted like young cells once again. Aging was no longer an immutable fact of life. Cell aging could now be reset at will. It was only in cells, not in patients, but it was the first time that cell aging was ever reversed, and it was the first important step toward clinical therapy.

The initial research was almost immediately followed by other work that went even further. The next step was to reverse aging in human tissues. It didn't take long to prove that the initial results were no fluke.

In the few years immediately following the turn of the century, several experiments—at Geron and in academic labs as well—showed that researchers could reverse aging both in cells and in the tissues made from those cells. For example, if you take the most common types of human skin cells (fibroblasts and keratinocytes) from an old person and allow them to grow together, these cells form skin tissue that is thin, friable, and typical of the skin we see in an old person. If you do the same with cells from a young person, the skin tissue that forms is thick, complex, and typical of the skin seen in a young person. But if you take skin cells from an old person and reset their telomere lengths, then the skin tissue that forms is typical of young human skin.[2] In short, we could reverse aging in old skin cells and thereby grow young skin.

---

[1] Shelton, D. N., et al. "Microarray Analysis of Replicative Senescence." 939–45.
[2] Funk, W. D. et al. "Telomerase Expression Restores Dermal Integrity to *In Vitro-Aged Fibroblasts in a Reconstituted Skin Model." *Experimental Cell Research* 258 (2000): 270–78.

Similar results occur with human vascular cells, using old cells to grow young vascular tissue[3] and using old human bone cells to grow young bone tissue.[4] In all cases, when we restore telomere lengths to the lengths seen in young cells, we can grow young cells from old cells: tissue that looks and functions like young tissue.

When You Reverse Aging in Cells, You Reverse Aging in Tissues

As we entered the twenty-first century, Geron had cloned the two key components of telomerase (together, hTERT and hTERC form the telomerase protein) *and* learned to use telomerase genes effectively. With these tools, Geron was already able to reset human telomeres in the laboratory, if not yet in the clinic. The key question became not whether we could reverse aging in *cells* and *tissues*, but whether we could reverse aging in entire living *organisms*. The stage seemed set for telomerase therapy to move from the laboratory to human trials. The theory was solid, the techniques were available, and I was ready to begin trying to help human patients.

Telomerase is the enzyme that extends telomeres and resets cell aging. It comprises two active parts: hTERT (human telomerase reverse transcriptase), which is the active protein part, and hTERC (human telomerase RNA component), which is the template part that tells the enzyme exactly which DNA bases to put into the telomere and in what order. hTERC acts as a blueprint, while hTERT does the work. Both together are necessary to extend human telomeres. These are the two parts of the enzyme that are absolutely essential for what we might call acute function (including therapeutic intervention), but there are dozens of other proteins and co-factors that are necessary for modulation, control, and long-term maintenance of the telomere. Moreover, it is almost impossible to

---

[3] Matsushita, S. et al. "eNOS Activity Is Reduced in Senescent Human Endothelial Cells ." *Circulation Research* 89 (2001): 793–98.

[4] Yudoh, K. et al. "Reconstituting Telomerase Activity Using the Telomerase Catalytic Subunit Prevents the Telomere Shorting and Replicative Senescence in Human Osteoblasts." *Journal of Bone and Mineral Research* 16 (2001): 1453–64.

keep up with advances, as new factors are elaborated upon almost every week in the current literature. Nor is there any sign that these advances are decelerating.

But even as the data began to show that the telomere theory of aging was correct, other factors were beginning to slow progress toward applying our knowledge to treatment of disease. While many people didn't understand telomere theory or its implications, that wasn't the problem. As so often happens when scientific advances languish, the problem lay not with the science but in human nature.

Two: Things Fall Apart

There's an old story about the first automobile to arrive in a small town. An old man looks at it and asks, "Where's the horse?" The driver says that it doesn't need a horse and explains about the gasoline engine. "Well, that's fine," says the old man, "but where's the horse?"

That's more or less what happened when Geron showed that telomeres were the engine of cell aging. It wasn't that the idea that aging can be reset was confusing—it was that it wasn't even heard. The idea that aging could be reversed was so obviously irrational that many people simply ignored the research and its implications. That was unfortunate, because the implications were profound: Not only could we reverse cell aging, but we could also reverse aging diseases.

In some cases, even those who were centrally involved in the research on telomeres and cell aging had a difficult time grasping the implications. Despite knowing the data and despite working closely with the key players, many at Geron Corporation, for example, had a very hard time believing that telomeres really played a role not just in cell aging but in human aging as well, let alone that telomere lengthening might be a desirable clinical intervention. As we will see, this problem—not believing the implications of the data—played a key role in preventing a rapid translation from the laboratory to the clinic.

Over the next decade, several biotech ventures fell by the wayside because the investors, initially eager to participate, had a hard time believing in the reality of what the research showed. When the prospect was simply a set of cellular experiments that might affect cell aging, and perhaps cancer, investors were plentiful, but when it came to clinical trials on human aging, investors had a hard time overcoming their assumptions that aging could never be altered. Like the old man staring at the first horseless carriage, they couldn't get around their lifelong assumptions.

There had to be a horse.

Unfortunately for patients, among those ignoring the clinical potential were the people making strategic decisions for Geron. The board of directors and the science advisors—although committed to Geron's focus on telomeres and cell aging—found themselves unable to change the assumption that aging was immutable. Despite their own research, they couldn't accept the idea of reversing aging, even in cells, let alone human disease caused by aging. The data was clear, but belief was far behind. Given that assumption and their understandable financial responsibilities for the success of the company, priorities began to shift.

At its founding, Geron's top priority had been to alter the aging process, but now two secondary goals took precedence, goals that were more credible and conservative. Cancer therapies and, to a lesser extent, stem cells became Geron's priorities. Aging was quietly shelved, a corporate embarrassment. Mike West, the key figure in creating the initial vision behind Geron, was first moved into a non-operational role, and then finally left the company altogether, leaving no one with a strong commitment to aging research and the potential clinical value of telomerase activation. After Mike West left Geron, all of the work and the patents were first licensed, then later sold off.

By 2002, Geron had already identified a number of strong telomerase activators, but these were felt to lack strategic value, and pharmaceutical development was shelved. A number of academic researchers took these compounds into additional testing, but most of this work was aimed at basic science, and very little at human disease.

Telomere research increased our basic knowledge of the biology and chemistry of telomerase, culminating in the Nobel Prize in 2009 for Elizabeth Blackburn, Carol Greider, and Jack Szostak, but this was purely academic work. There was almost no interest in the medical potential, let alone in moving telomerase to human trials, with the exceptions of a few entrepreneurs I'll discuss presently.

Geron not only stopped being the most promising leader in therapies for age-related disease, but actually applied the brakes on clinical progress. It held and defended the key patents, but it wasn't *using* them. This made it more difficult for new biotech companies to translate the science from basic telomere *theory* to clinical telomerase *therapy*. Many researchers and biotech companies considered moving into the void, but were hampered by Geron's hold on the patents. Geron had a claim over any product that a new biotech firm might create, making it almost impossible for new startups to get investors. Nonprofit academic research continued, but given the patent problems, most biotech investors avoided telomerase-related therapies despite their clinical promise.

Fortunately, Geron did eventually make their data and patents on telomerase activators available to others. In 2002, they sold the nutriceutical rights to these compounds to TA Sciences, and in 2011 sold that company exclusive rights to *all* of these compounds, nutriceutical or otherwise. Geron hung on to the stem cell patents a bit longer, but even these were sold in 2013 to a new biotech company, BioTime, whose CEO and founder is none other than Mike West. The stem cell work had finally returned to the very person who had started the field more than a decade before. Mike's company is working on developing embryonic stem cell therapies for age-related macular degeneration and for spinal cord injuries.

Geron now remains focused solely on telomerase inhibitors, arguably the least promising of their original technologies.

Noel Patton is the far-sighted businessman who saw the clinical potential of telomerase and bought Geron's telomerase activation technology. In the 1980s, Patton had been one of the first American small businessmen to venture into China. He had a deep interest in aging and as an early investor in Geron had kept a close eye on its patents. Geron had a set of four telomerase activators, steroidal molecules collectively called astragalosides. These compounds were extracted from the root of a plant—*Astragalus membranaceus*—that had been used as a tea in Chinese traditional medicine for more than a thousand years.

Patton recognized that the long history of human use qualified these botanicals as "presumed safe," so that they could be sold as "nutriceuticals." This designation doesn't allow therapeutic claims for human disease, although, as we will see, there is growing evidence that telomerase activation has significant clinical benefits for age-related disease. Marketers of nutriceutical products can, however, make other broad claims, such as their value in slowing aging, helping improve immune function, or improving general health and well-being. After buying the rights from Geron in 2002, Patton founded TA Sciences (TA for telomerase activation) and developed a way to extract and purify the astragalosides. By 2006, he was producing TA-65 gelcaps for sale in the United States.

ASTRAGALUS: AN ANCIENT HERBAL REMEDY

Astragalus membranaceus (also called *A. propinquinus*) is a perennial plant with hairy stems and small, symmetrical leaflets, whose appearance is similar to a common vetch. Growing approximately one half to one meter in height, it is native to northeast China, Mongolia, and Korea, but can be grown in most temperate areas, including in much of North America. Seeds are readily available on the Internet.

The dried root is typically harvested from the fully grown four-year-old plant. The root is dried, ground, and used as medicinal tea. In modern proprietary use as a telomerase activator, highly purified extracts from the ground root are made into commercial gel caps. Dried astragalus root is available in traditional Chinese herbal stores and teas, and extracts are commonly available wherever herbal supplements are sold. However, these are not reliable sources of the astragaloside molecules that activate telomerase. The few assays that have so far been done on commercial teas and extracts have found only trace amounts. *Caveat emptor.*

I once asked Patton why he got involved with telomerase activators. He grinned and said, "Well it certainly wasn't the money. I had been reasonably successful in my prior business and didn't need more money to maintain my lifestyle. It took eight years of losses and a lot of effort before TA Sciences could simply break even. And it wasn't because I was trying to save the human race, either. I was into my fifties when I first found out about telomeres and telomerase activators. I knew I wasn't going to live forever, but I didn't want to be unhealthy and dysfunctional as I continued to age. I like to ski, play tennis, and go dancing and wanted to keep right on enjoying my life. I suppose you could say, to be honest, that I did it to save my own ass."

If Patton had merely sold yet another product claiming to reverse aging, it would have been both unremarkable and not particularly useful. Patton went further. The cost of TA-65 is relatively high compared to products making similar claims, but the revenues helped fund clinical studies of the effects on aging. Patients had blood tests and physical exams aimed at identifying age-related changes such as immune function, cognitive function, bone density, blood pressure, visual contrast perception, skin elasticity, joint function, and so forth. The intent was to see if telomerase activation actually had the clinical benefits that many of us predicted.

TA Sciences was not alone, however. There were other efforts underway to bring telomerase to human trials and demonstrate the potential for curing disease.

Early in 2003, I was serving as the executive director of the American Aging Association when a wealthy, philanthropic couple flew me to California and offered me more than a billion dollars—carte blanche—to conduct clinical research on telomerase. They had read *Reversing Human Aging* and believed in the potential for treating human disease. I would have the resources to take the general telomere theory of aging from the laboratory to clinical trials. I called Cal Harley and discussed how we could now test telomerase in a medical setting. We were not alone: Many of our colleagues in both science and medicine knew the potential and were eager to push ahead with us, so I had strong support for my

Astragaloside IV

Cycloastragenol

Astragenol

Astragaloside IV 16-one

Astragalus molecules.
Cycloastragenol, perhaps the most active of these compounds, is also called TAT2. There are other potential telomerase activators, including GRN 510, AGS-499, and similar compounds.

project. I planned to test telomerase in human knees to cure osteo-arthritis, in human coronary arteries to cure atherosclerosis, and even planned on trials of telomerase to cure Alzheimer's disease. The project had the technical and medical expertise, and now had the financial support we needed. At the last minute, however, the night before we were to sign the financial documents, the donors argued among themselves, and the project ended, suddenly and permanently.

Once the offer to fund my clinical trials was withdrawn, the field settled into a slow siege. I wrote the first (and still the only) medical textbook on telomerase, *Cells, Aging, and Human Disease.*[5] While most telomerase research remained merely academic, there was a small contingent of biotech researchers and entrepreneurs who knew the medical potential and moved ahead. A number of small biotech startups were founded, each pursuing different approaches to re-lengthening human telomeres as a clinical intervention.

Years earlier, I had sat down with Bill Andrews (originally the director of molecular biology at Geron) at a conference in Italy and talked at length regarding the general telomere theory of aging. I not only convinced him, but Bill soon became the leading researcher in the clinical use of telomerase activation. In 2003, Bill founded Sierra Sciences, in Reno, Nevada, and focused on high-speed random screening of compounds, looking for better activators. Despite problems with investors and later financial issues, Bill persevered and finally identified more than 900 potentially valuable telomerase activators. Some of these were limited by toxicity or side effects, and even the best candidates initially showed only about 6 percent of the activity required to immortalize (i.e., reverse aging in) normal human cells. Still, using these compounds as the starting point, Bill and his team were able to design more effective compounds: Within two or three months they had compounds with both low toxicity and 16 percent activity and hoped that 100 percent efficacy might yet be within their reach.

Just as with Geron, however, it was hard to find investors who understood and believed in the clinical promise. Sierra's finances

---

[5] New York: Oxford University Press, 2004.

also suffered during the general financial crisis of 2008, making it difficult to continue with the research or to take any of their work to clinical testing. Bill began speaking tours, trying to increase public awareness of the science and its clinical promise as well as hoping to find new investors. The search for telomerase activators has been an obsession for Bill and his team, and doubtless Sierra Sciences will continue the search regardless of the obstacles.

While TA Sciences focused on bringing telomerase activators to market as nutriceuticals and Sierra Sciences searched for better telomerase activators, another group, led by researcher Barry Flanary, took a different approach, trying to find a way to administer the telomerase protein directly. In 2005, at Phoenix Biomolecular, he tried to use a new technology to deliver telomerase protein into cells. Hopes were high, but a number of obstacles—predominantly business and financial problems—finally forced Phoenix to close their doors despite the technical progress and the clinical promise of this approach.

As the end of the decade approached, the only practical result appeared to be the clinical research funded by TA Sciences. They now had data from hundreds of patients who had been taking oral telomerase activators starting in 2007. The first paper was published on this clinical data in 2011[6] and a second two years later.[7] Both papers looked for measurable changes in telomere lengths in white blood cells, as well as looking for evidence of actual improvement in clinical biomarkers such as in immune function or blood pressure as the patients aged. The 2011 paper showed that immune function could indeed be reset (i.e., made more like that of a younger person) by using the oral telomerase activator TA-65. The 2013 paper showed that cholesterol, HDL, glucose, and insulin levels could likewise be reset. These results were remarkable enough, but fell short of showing dramatic rejuvenation effects, prompting

---

[6] Harley, C. B. et al. "A Natural Product Telomerase Activator as Part of a Health Maintenance Program." *Rejuvenation Research* 14 (2011): 45–56.

[7] Harley, C. B. et al. "A Natural Product Telomerase Activator as Part of a Health Maintenance Program: Metabolic and Cardiovascular Response." *Rejuvenation Research* 16 (2013): 386–95.

many of us to want to find more effective ways of re-lengthening human telomeres.

The first decade of the new century saw some progress, as well as many setbacks in commercial development of age-reversing therapies. But there was a good deal of promising work being done in academic labs throughout the world. Most research was focused merely on the basic science—including the Nobel Prize–winning work of Blackburn, Greider, and Szostak. The more practical (and important) work was being done by those who saw the clinical potential of this field. These included Rita Effros at UCLA, who worked on immune aging and telomerase activators; Ron DePinho, then at Harvard, who showed that he could essentially reverse aging in certain genetically modified animals; and Maria Blasco at the Spanish National Cancer Research Centre in Madrid, who showed that she could also reverse many aspects of aging in several animal species.

As a whole, the academic and commercial progress was beginning to show—even to skeptics—the potential of using telomerase, yet the progress was frustratingly slow. On the other hand, there was a new generation of scientists who found it natural and reasonable to see telomeres as central to the aging process. And the public was slowly beginning to believe in the role of telomeres and the benefits of telomerase—even if those beliefs were often

ASTRAGALUS: CAVEAT EMPTOR ONCE MORE

The use of the astragalosides—as extracted from *Astragalus membranaceus*—was patented by Geron Corporation in 2000 and licensed exclusively to TA Sciences in 2002. Despite this restriction, several alternative sources—legal or not, reliable or not—have sprung up on the Internet claiming to offer astragalosides for use as telomerase activators. The legality, source, and purity of these compounds have been disputed, and their efficacy as telomerase activators is difficult to assess or prove. The controversy and the claims have made the market confusing and difficult for both the consumer and the clinician.

erroneous and overdramatic. Websites, television, spas, and various commercial enterprises claimed to have the latest information on how herbs, meditation, diet, pills, and other purportedly effective interventions might affect telomere length. Many of these interventions were openly advertised as able to re-lengthen telomeres, with the clear assumption that telomere shortening caused human aging. On the whole, many of these products would prove ineffective, while others were only minimally effective. Even the most effective known compound—cycloastragenol—wasn't nearly as effective most of us would have liked.

Although TA Science's oral formulation TA-65 was the only telomerase activator commercially available in 2013, several companies were considering skin creams, veterinary products, or medical (as opposed to nutriceutical) products based on various telomerase activators.

Three: Relaunch

> Great ideas pivot on the Janus'd cusp of history: looking forward they are obviously foolish, looking backwards they appear foolishly obvious. We are doubly blind.[8]

If the 1990s were a decade of hope and the 2000s a retrenchment, then the 2010s opened as a time for new beginnings. The public was gradually coming to understand that aging itself might be mutable and that telomeres played a central role. An increasing portion of the public was looking for ways to reverse their aging—by lengthening telomeres—and an increasing number of commercial companies were trying to meet this demand. There was at least one product that had demonstrated activity, but there were also several companies that offered the ability to measure telomeres. These companies had grown from the academic research labs and—much like the growing number of companies offering to identify patients' genes and mutations—they offered the ability to measure your age in terms of telomere shortening.

---

[8] Fossel M., Cells, *Aging, and Human Disease* (Oxford University Press, 2004).

The first of these was Telomere Diagnostics, founded by Cal Harley and based in Menlo Park, California. The second was Life Length, founded by Maria Blasco and based in Madrid, Spain. Using different approaches, both offered to measure telomere lengths and provide an indication of aging and the risk of disease. While these two companies had potential within the clinical market—hospitals and physician's offices, for example—they also were a good indicator of the growing interest and belief in the importance of telomere lengths in human aging and disease.[9] In addition, the existence of companies and labs that could measure telomere length made it increasingly easy to perform the research needed to begin human trials aimed at re-lengthening human telomeres as a treatment for age-related disease. Suddenly, there was a rapidly growing interest in developing a practical intervention that would use telomerase to reset gene expression and cure aging diseases.

Even the academic literature—once constrained to the narrow details—began to shift. More and more articles debated the value of measuring telomeres or focused on what we could do to lengthen them through diet, meditation, or supplements. A more fundamental shift had also begun: Those working with telomerase were starting to use it to change the aging process or to treat age-related diseases in animals. Telomerase was finally being recognized for its potential in clinical medicine. After all, if we can reverse age-related degeneration in a rat, then why not in a human patient with Alzheimer's disease? Yet even those doing the work on animals were reluctant to talk openly about the potential of telomerase for treating human disease.

Although telomerase activators and telomerase protein had been considered, and although the astragalosides were already in informal clinical trials, no one had yet taken the bold step and moved to human clinical trials. By 2010, a number of methods were available to deliver telomerase, including adenoviruses and liposomes. Adenoviruses had been used successfully, notably by

---

[9] Fossel, M. "Use of Telomere Length as a Biomarker for Aging and Age-Related Disease." *Current Translational Geriatrics and Experimental Gerontology Reports* 1 (2012): 121–27.

Maria Blasco in Madrid.[10] Liposomes were known to have problems getting into the cells, but might also work. The problem with using manmade liposomes was that it was hard to convince the body to let them into normal cells or to get them to cross the blood-brain barrier, a common problem in pharmacology.

In 2013, some people who had been involved in the defunct biotech company Phoenix Biomolecular attempted to bring Cal Harley and me into a project. They wanted to use liposomes to deliver telomerase genes, much as I had suggested twenty years ago. I argued strenuously for taking the technology not to the cosmetic market, but to trials against Alzheimer's disease. Unfortunately, a workable business structure was never created, and, despite the potential, the effort foundered from faulty execution.

Successful ventures require not merely funding and business talent, but a clear understanding of reality. As this chapter has stressed repeatedly, the major reason that telomerase hasn't yet moved into clinical testing over the past twenty years is that many of those involved—investors, management, and researchers alike—have a hard time getting their minds around the conceptual changes. Outside of a few medical articles and one textbook, the telomere theory of aging has rarely been explained, so even researchers are prone to misunderstandings. The key problem remains: People find it almost impossible to really believe that aging can be reversed. The notion of investing in a biotech company that aims to reverse human aging fails immediately. Biotech startup teams that begin their introduction to a group of venture capitalists by telling them we can reverse aging, rather than the cure aging diseases, fail before they begin.

Can we take telomerase to human trials?

Yes, but only by using tact, patience, and data. As I write this, we are now at the cusp of large-scale human trials that might use any of several approaches to re-lengthening telomeres in human patients: telomerase activation, telomerase genes, telomerase RNA,

---

[10] de Jesus, B. et al. "Telomerase Gene Therapy in Adult and Old Mice Delays Aging and Increases Longevity Without Increasing Cancer." *EMBO Molecular Medicine* 4 (2012): 1–14.

or telomerase proteins. The obstacles to curing age-related diseases such as Alzheimer's dementia are no longer technical. They have only to do with our ability to organize and carry out the few steps needed to bring the current work from human cells to human trials.

Summary

Over the past twenty-five years, there have been two distinct threads in the telomere field: the basic science and the clinical potential. The first area—the basic bench science—has received most of the headlines (and the Nobel Prize), yet despite its advances, this work has little significance in the lives of ordinary people. The second area—the ability to cure human disease—has only recently begun to achieve recognition, yet this is the area of truly historic significance.

The basic science began with the observation that cell aging was linked to changes in telomere lengths. This view—the limited telomere theory of aging—gained acceptance within a decade or two of its formulation. The ability to cure disease, however, depends on the notion that although telomeres drive cell aging, it is cell aging that drives human aging and age-related disease. This broader view—the general telomere theory of aging—was first expressed twenty years ago, and is now gaining acceptance. Clinical advances have been delayed by our inability to grasp the concepts, but in the past several years, things have begun moving ahead again. Scientific and public understanding are both growing as we begin to work on what will become the greatest medical breakthrough we have ever known.

We are on the cusp of curing aging and its diseases.

~~~

CHAPTER FIVE

Direct Aging:
Avalanche Effects

When we think of aging, we stop thinking.

We simply glide over the idea of aging and focus on the diseases associated with it. Some of those are curable, as with certain cancers. For others, we can at best treat only the symptoms. As for aging itself, we can choose to go gentle into that good night or to rage against the dying of the light, but, either way, it has always seemed inexorable.

Because, until recently, we've never understood how aging occurs; we've assumed it is an inalterable fact of life. And so, our medical approaches to aging itself have been nothing more than palliative. Our minds have been closed to anything more.

How astonishingly different this is from the way we see all other diseases!

Infectious diseases inspire an entirely different response: What can we do to cure or prevent infection, to make people healthy again? We've invented immunizations that have all but permanently defeated diseases such as smallpox and polio. We've developed antibiotics, antivirals, antifungal agents, and novel approaches to sepsis. We've even unraveled the genomes of infectious agents. What will be next, we wonder? Even as we worry—appropriately—about evolving resistance to antibiotics, our responses remain optimistic, dynamic, and aimed at innovation.

None of this arises when the notion of age-related disease enters our minds. We accept aging passively, without question, in silence.

> Our lives begin to end the day we become silent about things that matter.
>
> — Martin Luther King Jr.

It is time to change our lives before they end, because the diseases of aging matter to every one of us. To do that, we must understand those diseases that eat away at our lives. We need to know how aging works—as articulated in Chapter Two—and a fluency in how genetics and telomeres combine to result in the diseases that we plan to cure. As you will see, our genes don't operate alone, nor are they necessarily our fate. Our genes themselves are unchanging and influential, often hidden in their complexity and purpose. But our genes *do* change their patterns of expression in response to changes in our telomeres and in our environment and our behaviors. Neither telomeres nor behaviors can alter our genes, but gene expression is variable and responds to everything that happens to you, to your tissues, cells, and telomeres.

Genes are often thought of as directive and all-powerful, that "genes cause disease," which leads people to ask: *Which* genes cause *which* diseases? This is no more accurate than thinking of telomeres as causing aging: Genes are associated with disease, and are sometimes causal, but disease is almost never simple in regard to cause and effect. When we look at most of the diseases of aging, genes don't "cause" those diseases, nor does telomere shortening "cause" the aging process.

The reality is much more subtle.

To put it succinctly: *Age-related diseases occur when telomere shortening exposes our genetic flaws.* To understand the relationship between genes and aging—particularly the diseases of aging—let's turn back to the analogy in Chapter Two, in which we compared aging—the result of shortening telomeres—to sailing on a large body of water in which the water level keeps gradually dropping. The lower it drops, the more likely our ship is to strike a rock or run aground on a shoal. In time, this body of water becomes unnavigable. When we are young and our telomeres long, we are in no danger of striking these hazards. As our telomeres shorten, we're in increasing danger of foundering on these now-exposed rocks. Eventually, it happens to all of us.

This is the actual relationship between genetic predilections and diseases of aging. The same is largely true of behavioral risks and diseases of aging. A gene that increases your risk of heart disease doesn't manifest itself in atherosclerosis when you are five years old, but might become fatal by the time you reach fifty. Similarly, lack of exercise, poor dietary choices, and tobacco use may not result in a heart attack until you begin to grow old. It is probably not the only duration of exposure to these risks, but the gradual erosion of your telomeres (often accelerated by those same risks) that exposes the diseases of aging.

When we consider genes that correlate with the diseases of aging, such as APO-E4 with Alzheimer's disease or the genes related to cholesterol metabolism with atherosclerosis, there is never 100 percent penetrance (i.e., some people have the gene but not the disease, while others have the disease but not the gene). Yet the simplistic assumption is that if we could only locate the full panel of the genes that "cause" the disease, then we could predict the disease with certainty. The reality is that it is not the *genes* that cause the disease, but their *expression*, and gene expression is controlled by myriad factors including both our behaviors and our telomeres.

Genes cause disease depending on *how they are expressed* and in *what circumstances.*

A "dangerous gene" or a gene that "causes disease" is no problem if it isn't expressed sufficiently or is expressed only in appropriate circumstances. The circumstances include your diet, your behaviors, your environment, your other genes, and your age. Genes that were perfectly benign when you were young may be fatal as you get old.

As telomeres shorten with age, a large number of genes change their patterns of expression. Some increase their expression, some decrease their expression, many change how they respond to other genes or to changes in the environment. If we believe that diseases such as Alzheimer's and atherosclerosis are simply caused by specific genes whose effects accumulate over time, then we must conclude that there is nothing to be done about age-related diseases (short of altering our genes). If, however, we recognize the complex reality—that the changing gene expression is a result of telomere shortening—then we conclude that we *can* do a great deal about age-related disease.

Understanding how aging causes disease tells us how to cure disease.

If age-related disease results from changes in gene expression that result from telomere shortening, then if we re-lengthen those telomeres and reset gene expression, we may cure the diseases of aging. To return to our metaphor, if we raise the water level, those rocks and shoals are no longer hazardous, and we can once again navigate our lives in safety.

For example, let's consider a simple age-related problem, varicose veins. These are commonly assumed to be the result of gravity multiplied by decades. If some people are more likely to get them than others, we assume that's genetic variability. The varicosities accumulate and—short of surgery, which is mostly cosmetic—little can be done to intervene. But what if the varicosities are *not* simply a function of time and gravity, but of gradual changes in gene expression? What if it isn't the years that cause them but the accumulation of poor cell repair? If that is the case, then resetting the pattern of gene expression might well enable the tissues to repair the damage. We can't roll back the years, but telomere theory opens the door to rolling back physical aging.

Today we are looking at a collision between a long-held assumption and a new insight.

The assumption—shared by the public, scientists, and researchers—has been that aging is a simple, passive accumulation of damage for which there is no realistic intervention. Aging can't be reversed and age-related diseases can't be cured. Age-related diseases can only be endured or, at best, treated symptomatically or cosmetically. You can't change your genes, nor can you avoid the passage of time. We can cure or prevent many infectious diseases, but age-related diseases are everyone's fate. *Que sera, sera.*

With most pivotal advances in human history, the major obstacle has been the assumption that change was impossible. Such assumptions are always self-fulfilling. We only make progress when thoughtless assumptions are shattered by thoughtful insight. In this case, the insight is that aging and its diseases are the complex, dynamic result of gradual changes of gene expression, the effects of which are largely reversible, and that telomere elongation is an efficient point of intervention in aging and age-related diseases.

In this chapter, we will focus on *direct* age-related diseases—those in which the cell that ages is that cell that shows the pathology. In the next chapter, we will consider *indirect* age-related diseases—those in which one set of aging cells is responsible for pathology in other, normally non-aging cells, "innocent bystanders," as it were.

Direct aging disease is an "avalanche" of cell pathology that occurs when cell aging disrupts cell function. One example of direct aging is osteoarthritis—which we will consider in more detail below—in which the cells that line the knee joint, for example, slowly lose telomere lengths, change gene expression, and become dysfunctional, causing a gradual loss of joint surface, along with pain and disability. The cells that line the joints—the chondrocytes—undergo direct aging, and these are precisely the cells that fail, causing arthritis.

Let's use a model to understand how cell aging causes a direct age-related disease. We'll invent a cell, put in a gene or two, add a protein or so, and see what happens as the cell undergoes aging. We will make the cell, and our discussion, quite unrealistically simple, for the sake of example and clarity.

There is a gene, typically thought to play a role in aging, that is responsible for the enzyme superoxide dismutase (SOD). (Actually, SOD is a family of several different enzymes, but we'll think of it as one.) SODs are critical in scavenging free radicals that escape your mitochondria and damage your cells.

So our cell has several players: a telomere, a SOD gene, the SOD itself, free radicals, and a single type of molecule that is the cell's "main product," in this case, a protein needed to build and maintain muscles.

Let's say our young cell has a pool of 100 SOD molecules and 100 protein molecules. These pools are *dynamic,* because every day our imaginary cell creates fifty brand-new SOD molecules (anabolism) and breaks down and recycles fifty SOD molecules (catabolism). The same thing occurs in our pool of protein molecules. Both pools of molecules are always exactly the same *size,* but the specific molecules in that pool are always changing—different molecules but always 100 of them in each pool. Because the breakdown is random, about half of all the SOD molecules were created today, while about half of the SOD molecules may be a bit older—although not much older. The pool of protein molecules is doing precisely the same thing: Half are new, half are a bit older.

Unfortunately, this being a typical living cell, there are lots of free radicals that are randomly damaging any molecule they can find. We'll assume that there are just enough free radicals around that every day they damage about 1 percent of the molecules in the cell. Of course, these free radicals would damage a lot more of our molecules if it weren't for our "policemen," the SODs, which are busy "arresting" free radicals and making sure they don't create even more mischief.

We can establish a formula that tells us the percent of damage in our cell (M is the metabolic turnover rate, which slows with age):

$$x = 1 + [x(100\% - M)/100]$$

- In our *young* cell, we replace 50 percent (0.50) of the molecules per day (M), while free radicals damage 1 percent of the molecules per day. The percentage (x) of damaged molecules in the pool is 2 percent.

- But in an *old* cell, we replace only 2 percent (0.02) of the molecules per day (M), while free radicals still damage 1 percent of the molecules per day. The percentage (x) of damaged molecules in the pool now rises to 50 percent.

So in our young cell, about 2 percent of SOD molecules simply don't work and about 2 percent of our protein molecules are damaged as well, which is normal. The size of the molecular pool and the rate of metabolic turnover are good enough to deal with the damage with the least wasted energy. Young cells have a high metabolic rate (they use a lot of energy), a high rate of turnover, and few enough damaged molecules that they don't matter.

In the old cell, however, the telomeres have shortened so that the rate of expression is "turned down" for both the SOD and protein genes. The result is that the rate of metabolic turnover is slower. Instead of replacing 50 molecules per day (SOD and protein), our old cell replaces only 2 molecules per day. The cell has the same number of molecules, but the rate of turnover is much less. Molecules damaged by free radicals now "stick around" longer, so that the proportion of molecules that don't work increases from 2 percent to 50 percent. This isn't because we have more free radicals

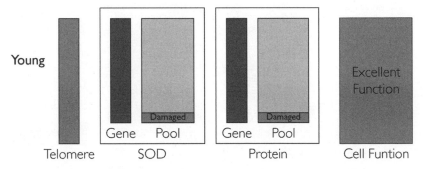

In young cells, telomeres are long, gene expression high, damage low, and the cell has full function.

or because the damaged molecules are never replaced; the problem is that we don't replace the damaged molecules as fast as we did in the young cell.

It's actually worse than that. Because the SOD molecules protect our cell's protein molecules against free-radical damage, the damage rate for protein molecules increases by more than 50 percent, perhaps to 80 percent. And it gets worse yet. Because the SOD molecules are more likely to be damaged, they can't even protect *themselves* against damage, so the SOD molecular pool also shows more damage than the formula would suggest. It's a vicious cycle. The formula assumes a constant rate of damage (1 percent per day), but now the rate of damage is climbing just as the rate of replacement is going down. So our cell's protein molecules are even more damaged than we thought. Instead of only 80 percent, perhaps 90 percent are damaged.

Yet all we did was slow the rate of gene expression.

We didn't increase the actual number of free radicals that the cell produced, nor is the damage permanent. It's just that we are no longer replacing the damage as quickly as in the young cell. While these figures (the rates of repair, for example) are merely examples to show the effect, gene expression does slow with age for many proteins, and the overall effect is quite real. In actual cells, moreover, the mitochondria themselves begin to create more

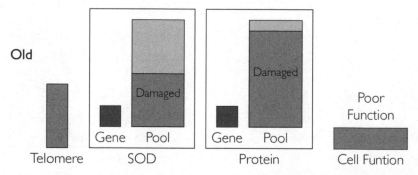

In old cells, telomeres are short, gene expression low, damage increases markedly, and cell function decreases.

free radicals, as well as leak more than they do in young cells, in both cases due to slower molecular turnover. Finally, free radicals are not the only source of molecular damage. In actual cells, the reality is infinitely more complex, but the effect of aging remains overwhelming: As telomeres shorten, cells become dysfunctional.

Shortening telomeres create an avalanche of dysfunction, ending in disease.

Human age-related disease occurs in exactly this way. In the case of direct aging diseases, the outcome can be described within a single cell type, such as chondrocytes, leukocytes, fibroblasts, and so forth. In each type of cell, aging results in a typical age-related disease. In the rest of this chapter, we will look at particular age-related diseases and specific organs that are the habitat for each of those diseases. The diseases of aging have one thing in common: They cannot currently be treated. Yet, in one sense this statement is untrue. For example, we can replace joints and bypass coronary arteries, and we can control cholesterol, serum glucose, and blood pressure. However, there is not a single age-related disease that we can cure, prevent, or even slow down very much by using current medical interventions. Not a one. As we look at each disease, we will sketch out the current therapies, and what we will see is that unless we can extend telomeres, the prospects are abysmal for patients, for caregivers, and for health care costs.

A Few Notes on the Discussion to Follow

While I want to give the reader a sense of the costs of age-related diseases—in terms of both human suffering and the cost of treatment—the numbers I give are very rough. It is all but impossible to construct reliable worldwide statistics. For example, it's easy to get good estimates on the number of Alzheimer's patients in the US, the UK, or Australia, but almost impossible to get the numbers for some countries in Africa or Eastern Europe. For that reason, most of my figures are taken from databases available in the United States.

Also, many of the figures are not well defined. For example, the medical diagnosis of Alzheimer's dementia has been undergoing substantive change, especially as new diagnostic biomarkers come into common use.

Finally, the financial costs are hard to quantify and have multiple definitions. Do we measure only insurance payments, hospital costs, government medical budget line items, or something else? Do we include only "direct costs" such as hospitalizations, surgeries, and medications, or do we include "indirect costs" such as the care given by unlicensed family members, the lost work opportunities, and other less tangible costs?

Nonetheless, rough figures are valuable even if we're only accurate by more or less than a few billion dollars or a million people who are suffering from the diseases of aging. Because the numbers are so huge, that's close enough. Nor does the scale really matter to those who suffer.

Unless we can truly intervene, unless we can extend human telomeres, there is one fact that remains distinct and inescapable: In the long run, these diseases are waiting for us all.

As for the diseases themselves, the key question is intervention. Can we prevent or cure disease? In every case, this issue is practical and compassionate, rather than academic or purely scientific. I am not concerned with whether you gain a deep scientific understanding of each age-related human disease, but rather that you come to grips with the human side of such diseases and that you understand how cell aging causes them and why telomerase therapy may cure them. There is now good reason to think of aging not as an inevitable mystery but as a set of specific changes, resulting in specific diseases, all of which might be mitigated or cured by telomerase therapy. We will explore age-related disease with prevention and cure foremost in our minds.

We will start with the immune system, our general defense against a wide gamut of diseases—infectious, malignant, and autoimmune—with far-reaching effects throughout the body. A large portion of the elderly succumb to infection or cancer, rather than to Alzheimer's disease, atherosclerosis, COPD, or other age-related pathology, even when these other age-related diseases are present. In this respect, immune-system aging is the weak link and becomes the final common denominator for so many deaths among the elderly.

We will then look at other diseases, with respect to aging cell types or organs. We will start with our joints and bones (osteoarthritis and osteoporosis), then our muscles, skin, hormones, lungs, gastrointestinal system, kidneys, sensory system, and others. Then, in the next chapter, we will look closely at the two most pressing diseases of aging: Alzheimer's disease and atherosclerosis.

The Immune System

The immune system is ubiquitous, ceaseless, and critical to survival.

As in the nervous system, immune function comprises both what we might call instincts and more complex, learned behaviors. Even at birth, your immune cells are perfectly capable of recognizing any number of external threats, but they become more and more capable and discerning with maturation and experience. With every invading virus or bacteria, with each fungus and potentially cancerous cell, your body becomes more discerning and adept at fighting off threats.

And as with the nervous system, even though its learning continues throughout life, it is balanced against the gradual and inexorable failure of memory. The young immune system is naïve but vigorous, while the old immune system is more knowledgeable, but slower and clumsier in its execution. Immune senescence is not a matter of the immune system not *recognizing* an invader— such as pneumococcal pneumonia—but a matter of *responding too slowly and erratically* to manage the infection before it topples the entire organism—as through death by overwhelming sepsis. The aging immune system is tragically reminiscent of an old joke about pathologists, who know everything and do everything, but too late to help the patient.

The cells of the immune system derive from stem cells in the marrow, as do red blood cells. The lineage divides into two main branches, the lymphoid cells and the myeloid cells. The lymphoid branch—named because these cells circulate in the lymphatic system as well as in the blood system—includes the natural killer cells (NKCs), T cells, and B cells. Together these are the lymphocytes responsible for most of your immune function. The myeloid branch includes thrombocytes (which play a role in clotting), erythrocytes (red cells, which carry oxygen), and a number of white cells (basophils, neutrophils, eosinophils, and macrophages), which are also part of the immune system.

Each of these types of cells has its own special function and its own characteristic patterns of behavior and cell division, which means that each component of the immune system ages in a slightly different manner. Therefore, the immune system not only fails with age, but fails in complex, surprising ways, rather than as a single unit.

The aging immune system doesn't grind to a halt; it flails, sputters, and blunders about your body with gradually ebbing effectiveness.

Immune senescence, though a common contributor to disease and death in the elderly, is seldom recognized and even more rarely diagnosed. The clinical manifestations—what we see and what we treat—are chronic inflammation, rheumatoid arthritis, autoimmune disease, increased risk for pneumonia, sepsis, cellulitis, shingles, and, in some cases, various forms of cancer. Age brings a slight decrease in peripheral white blood cell count, but not enough to increase the risk of infection. To the contrary, most elderly patients do mount a rapid response—elevated white cell count—when they have an infection. Moreover, many elderly patients have a higher than normal peripheral white cell count, and this often correlates with atherosclerosis; many patients with heart attacks and strokes have elevated white cell counts prior to the acute event. In short, the problem with aging immunity is not something as simple as having fewer (or more) circulating white cells, but the response of the immune system as a whole. Instead of having an accurate and precise response to infection, the elderly immune system might respond even when there is no infection—inappropriate and chronic inflammation—and might not respond when there is an actual infection. It is not that it can't respond, but that the response is often wrong, either too slow or poorly targeted.

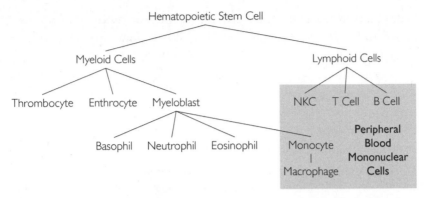

The various kinds of cells in the immune system, divided into blood cells and lymphatic cells, all of which originate from stem cells in the bone marrow.

Like other actively dividing cells in the body, immune cells show shorter telomeres with aging, although the pattern of that shortening is complex. For one thing, B lymphocytes divide more frequently in the marrow, prior to entering the circulation, while T lymphocytes divide more frequently after entering the circulation. In addition, only about one in thirty immune cells is actually in the circulatory system; the rest are in tissues, especially lymphatic tissues, and the pattern and timing of their entry into the circulatory system can vary between different cell types.

B lymphocytes have a specific pattern of cell aging. About 2 percent of the body's entire complement of B cells is created every day, although this rate falls substantially with age. The number of lymphocytes remains relatively constant, as cell division is balanced by death in lymph nodes and tissues. Even when the peripheral white-cell count rises, as in infections, it is more the result of existing cells entering the bloodstream than the result of cell division creating more lymphocytes. While the B cells derive originally from the stem cell compartment in the bone marrow, these cells are immature and require "editing;" most self-reactive B cells (those that might cause autoimmune disease) are removed before leaving the marrow. Those lymphocytes that do leave the marrow circulate continually until they either encounter their specific antigen—and become activated—or they die. Once they are in circulation, they continue to divide, especially within the spleen. The result is that about half of the cell division—and hence half of their cell aging—occurs *after* the B cells leave the marrow. The average telomere length of circulating B cells depends on the balance between relatively new cells and older cells. The newer cells have come from more recent stem cell divisions and hence have shorter telomeres); the older "memory" B cells, which have longer lives, have longer telomeres.

T lymphocytes, on the other hand, show an entirely different pattern. While B cells divide early in their course, then stop dividing when they are exposed to antigens and become memory B cells, T cells actually divide very infrequently at first—in the thymus—but once they are activated, they divide often. While the overall T cell pool is maintained at a constant size, the rate of cell division is high in the periphery. The result is that the T cells that have been in circulation the longest will have the shortest telomeres and the newest additions will have the longest telomeres.

Further, measuring telomere lengths in the peripheral lymphocytes can be misleading; as noted, they are only about 1/30 of the total number. Moreover, when we repeat measurements, we are seldom measuring the same cells. In one such measurement, we might be measuring cells that have just entered the circulation and hence accurately reflect the stem cell telomere lengths, while another measurement might be skewed if the cells have divided repeatedly in the periphery, leading us to underestimate telomere lengths. Estimates of immune system aging—or health status—based on peripheral telomere lengths are almost certainly useful, but must be carefully interpreted.

Other than telomere lengths per se, the key question in immune aging is, "How do the cells actually *function*?" Granting that changes in function are the result of changing patterns of gene expression, which result from gradual shortening of the telomeres, it is the functional changes that are key. Although most immune cells show such changes, the most obvious changes occur to the T cell population. These cells become "sloppy" (i.e., signal transduction becomes poorly regulated) and less capable of making critical cell products (e.g., lymphokines). With age, there are fewer naïve T cells and a growing inability to rapidly divide in response to infection and antigens. As with so many other systems in the body, the aging immune system responds far too well to some things, as occurs in autoimmune disease, and not well enough to others, as occurs with cancer cells, viruses, etc. While many cells become continually activated, causing chronic inflammation, others, such as the natural killer cells and other cytotoxic cells, become far less effective.

In addition, the gradual loss of telomere length within the stem cell compartment results in a decreased ability to replace hematopoietic cells, including erythrocytes, lymphocytes, and other cell types. Within the lymphocyte population, such cells have a gradually diminishing rate of turnover, so that an increasingly higher percentage of them don't function well. Within the erythrocyte population, the result can be a gradual anemia. Like the anemia of chronic disease, the anemia of advanced age, while rare, does not represent an exhaustion of the stem cells, but a relative failure of those cells to divide with sufficient frequency. In short, aging can bring on a relative age-related aplastic anemia as a result of bone marrow failure of the stem cell compartment.

THE AGING IMMUNE SYSTEM: QUICK FACTS

Age: Generally speaking, the older the adult patient, the more impaired the immune system.

Statistics: Data is hard to come by, because it is difficult to isolate immune senescence from other aging diseases. For instance, if an elderly woman stumbles due to poor eyesight, fails to catch herself from falling due to poor muscle strength, breaks her hip due to osteoporosis, has complications due to poor peripheral circulation, then gets an infection and dies, to what do we attribute the cause of death? For similar reasons, costs for immune senescence are difficult to isolate, but they are unarguably high.

continued on next page

Diagnosis: Distinct diagnoses of immune senescence are rarely made. In lieu of ordering unnecessary laboratory tests, physicians normally assume that elderly patients have impaired immune response.

Treatment: There is no therapy for immune senescence. Physicians counsel an adequate diet and routine immunizations (even though immunizations are less likely to generate an adequate immune response in the elderly than in young adults).

The clinical results bear out all these changes. As we age, we are more susceptible to infections, cancer, chronic inflammation, and autoimmune disease.

Osteoarthritis

Osteoarthritis is a disease of failing chondrocytes. These are the only kind of cell that resides in the cartilage of our joints, hidden like small seeds. Their job is to produce and maintain cartilage, the stiff, gelatin-like connective tissue made mostly of protein that forms the two slippery joint surfaces that rub across one another as the joint moves. Cartilage makes joint movement smooth and liquid, minimizing wear and tear and permitting rapid, efficient motion.

As in our example of key proteins made by an aging cell, the proteins of the cartilage—largely collagen and proteoglycans— are the critical products of the chondrocytes. These proteins are relatively stable, but still undergo recycling by chondrocytes, which degrade existing cartilage matrix and secrete new matrix to replace it. In short, there is a gradual but critical recycling of the joint surface, and it is this gradual recycling that slows with advancing age.

The result of this slowing is that, beginning in early middle age, the matrix gradually begins to accrue damage. The rate of damage itself is constant, due to the normal pressure and stress caused by movement of any joint, particularly in weight-bearing joints, such as the knees and hips, but the chondrocytes gradually lose their ability to keep up with this damage. As the chondrocytes age, their telomeres shorten, gene expression of the critical proteins slows down, the turnover of the proteins in the cartilage becomes slower and slower, and the cartilage begins to fail. As it does, the cartilage gets thinner and begins to fray, and the chondrocytes lose their physical protection from compression and shear forces, resulting in an increasingly rapid loss of chondrocytes. Unfortunately, telomere shortening also makes the chondrocytes less responsive to the need for cell replacement and less capable of dividing as well. The outcome is that chondrocytes are not only slow to replace the cartilage matrix, but there are fewer and fewer chondrocytes to do the job at all.

Oddly enough, the articular cartilage—and the chondrocytes that live within it—has no blood supply. It is nourished only through the synovial fluid. Oxygen and nutrients must diffuse from the distant capillaries through synovial fluid and cartilage to the cells; waste products from the cells must make the same journey in reverse. The active use of your joints—the movements of your daily life—helps this diffusion take place, so it is critical to the survival of the chondrocytes and hence to the survival of your cartilage. Yet even with optimal exercise, the telomeres gradually shorten, so that the chondrocytes no longer function adequately. New cells cannot arise from the blood vessels, but only from existing chondrocytes, which only accelerates telomere loss.

The more we ask of our chondrocytes—through excessive impacts, injuries, or high body weight—the more accelerated the cell aging and the earlier and more severe the osteoarthritis. The hand joints bear less weight than the knees, but incur more injury in occupations with repetitive movements, particularly with impacts or injury, such as in boxing. The vertebrae, hips, knees, and ankles are continually stressed by compressive forces across the joint surface, made worse again with repetitive impacts—runners, basketball, football, and soccer players are more likely to be afflicted, as is anyone with a job that involves repetitive or traumatic joint usage.

Overall, osteoarthritis is not caused by "aging" in the sense of increasing age itself, nor is it uniform, clearly predictable, or well-correlated with other age-related diseases. The onset and course of osteoarthritis results from telomere shortening, but is also the end result of factors that control telomere shortening in the affected cells, including genetic predilections, personal behaviors, diet quality and quantity, traumatic injuries, infections, and a host of other environmental factors. As ever, telomeres don't so much "cause" osteoarthritis; rather, they are the single common factor

OSTEOARTHRITIS: QUICK FACTS

Age: Onset generally begins between the ages of forty and eighty.

Statistics: OA is the most common form of arthritis, far more common than rheumatoid varieties. About 14 percent of adults have OA, but about a third of those aged sixty-five. The Centers for Disease Control estimates 27 million patients in the US have osteoarthritis. Risk factors include high body mass, joint injuries, and any activity (sports or occupation) that causes repeated impacts to a joint. Women have higher risk than men, particularly after menopause. Common sites are knees, hips, hands, feet, and spinal joints.

Cost: Annual US costs are more than $185 billion; $29 billion for knee and $14 billion for hip replacements, with work-related costs between $4 and $14 billion. OA is a frequent cause of disability. These costs are rising as people live longer and as incidence of obesity increases, both of which increase the frequency and severity of this disease.

Diagnosis: The symptoms alone—pain in the joints, often with swelling—are fairly reliable indications. Lab and radiology studies may rule out OA, but radiology (or more rarely CT, MRI, or arthroscopy) is generally used to confirm the disease.

Treatment: OA is usually addressed with pain medications, range-of-motion exercises, and avoidance of further joint impacts. However, none of these approaches has ever been shown to slow the progress of OA, nor have supplements like glucosamine sulfate. Current definitive therapy, total joint replacement, is chosen by about 5 percent of patients. While large joints (knees and hips) can be replaced, the disease itself cannot be stopped.

in an enormous and complex cascade of pathology that results in this disease. And for this reason telomeres are a more effective and efficient point of clinical intervention than any of the other factors that also play a role.

Osteoporosis

Osteoporosis is the gradual weakening of bones with age. Most people understand the term to mean that the bone becomes more porous, which is accurate, although it doesn't adequately convey the severe weakness that is the actual clinical problem. Osteoporotic bone breaks quite easily. There are endless true stories of elderly patients who have broken a bone—a lumbar vertebra or a rib, for example—merely by sitting down too hard or by coughing too strenuously. In a healthy young adult, a fractured femur is the result of severe trauma, like an automobile crash, while an elderly patient with osteoporosis might suffer such an injury merely by falling from a standing position onto a carpeted floor. Although very few patients ever die of osteoporosis, a great many suffer from unexpected and painful fractures, and many elderly patients die of its complications, such as pneumonia, clots, sepsis, etc.

Osteoporosis is the single most common bone disease and, like many other age-related diseases, would probably be universal in all elderly patients if they didn't die from other age-related diseases first. Most people mistakenly think of osteoporosis as being due to having too little calcium. In fact, we might say that the body has quite sufficient calcium, but in all the wrong places. The elderly body, for example, may have little calcium in the bones, but still have far too much calcium as deposits in the coronary arteries. It would be more accurate to say that it is caused by a lack of the protein scaffolding—the matrix that binds the calcium and other mineral components of healthy bone, such as phosphorus. This observation is borne out by the bulk of clinical studies showing that increases in dietary calcium and other simplistic dietary approaches have little effect on the progress of the disease.

It's not how much calcium you have, it's where your body puts it.

It is certainly true that those patients who have too little calcium in their diets would benefit from ingesting more prior to the onset of osteoporosis, but once a patient has clinical osteoporosis, increases in dietary calcium have little or no clinical benefit.

Of course, mere calcium intake isn't the issue. It's the complex interplay of calcium, vitamins, and hormones. Those with a life-long vitamin D deficit, for example, will also have deficient bone growth and poor bone maintenance, but this is neither osteoporosis nor its cause. Osteoporosis is not a dearth of calcium, nor can it be treated by supplementation alone. On the other hand, women who have had multiple pregnancies—and hence a recurrent draw on their own calcium supplies as they grow fetal bone during those pregnancies—are at higher risk, as are women in general, particularly post-menopause, when estrogen levels fall. Nevertheless, none of these problems speaks to the cause of osteoporosis, and none of the associated simple dietary remedies or supplemental estrogen or vitamins has ever been shown to stop or reverse the progress of the disease.

Bone maintenance—and hence osteoporosis—involves at least two types of bone cells: osteoblasts, which build bone, and osteo-clasts, which break it back down again. One might reasonably wonder why the body doesn't just build good bone and be done with it. The answer can be seen in two ways. The first is the same answer that we used to explain molecular turnover within cells: The body is constantly recycling in order to ensure that molecules—and in this case, bone—don't accumulate a high rate of damage over time. It's much like the process of continual upkeep on a house, in which all of the parts are routinely replaced, fixed, painted, and repaired, with the result that even an old house can be kept in good solid condition indefinitely. The second way of looking at the problem is simply to ask what happens when you break a bone; the body remodels and heals the fracture. This healing involves getting rid of the damaged bone and replacing it with normal bone. Actually, it's probable that micro-fractures occur moment to moment, even during the most marginal daily activity, and that a process of con-tinual recycling repairs these micro-fractures in the bony matrix just as it does major fractures and other kinds of damage.

This recycling of the bony matrix slows with age, just like the molecular recycling in the model cell I described earlier. The result is the same as well: The slower the recycling process, the greater the percentage of accrued damage and the more likely we are to undergo catastrophic failure. In the case of osteoporosis, the slower the recycling, the greater the percentage of weakened bone and the more likely we are to suffer fractures. Moreover, there is another, separate problem in aging bones: a growing imbalance between bony destruction (osteoclasts) and bony growth (osteoblasts).

The bone itself might be seen as a flexible and complex web of ropes, which the body coats with a tough, inflexible layer of concrete. The ropes are the protein matrix that makes our bone strong, the concrete is the calcium and phosphorus that makes the bone hard and durable. During the initial stages of healing (also in the growing fetus), bone is laid down in a "woven" form that is weak and pliable, but can be rapidly formed. Afterward, as healing becomes complete ("bony substitution"), bone is remodeled into a "lamellar" form that has greater mechanical strength and is more resilient and longer lasting. In the initial "woven" form, the smaller number of collagen fibers appear to be haphazard and random. In the lamellar form, the much more numerous collagen fibers are laid parallel in sheets, each set of fibers at right angle to the next set. This construction is similar to that of plywood and has similar advantages: enormous strength and resistance to damage.

This same process occurs in the process called remodeling, which takes place during growth and as a result of exercise and changing patterns of bony stress, as when an athlete begins a new sport or activity and the bone responds appropriately. However, bony remodeling is also a continuous process, even when full growth has been complete and there is no change in physical activity. While this is partly in response to micro-fractures that occur during daily activity, it occurs even in the absence of any damage. Bone is constantly undergoing a dynamic process of remodeling, of resorption and regrowth, even if the bone maintains the same shape and function.

In the young adult, the constant recycling of bone—the breaking down and building up—is kept in almost perfect balance. This process maintains bone at its optimal strength and size, while providing a ready repository for calcium and phosphorus for the rest of the body. In the average adult, about 10 percent of bone is remodeled—broken down and regrown—every year. But with age, the remodeling rate falls *and* the rate of replacement declines faster than the rate of destruction. This accounts for both the gradual loss of bone and the fact that bone fractures heal more slowly in the elderly.

Bone turnover is promoted by certain endocrines—growth hormone, thyroid hormone, estrogens and androgens—but there is no evidence that osteoporosis is a result of hormone levels diminishing with age. These and other hormones increase the osteoblast secretion of cytokines that increase bony resorption by stimulating the osteoclasts and promoting the development of new osteocytes from stem cells. Osteoclasts increase their resorption when stimulated by parathyroid hormone and vitamin D and indirectly by an increase in several cytokines (RANK-ligand and interleukin 6). Bone resorption by osteoclasts is inhibited by osteoprotegerin and calcitonin. Note that some endocrine influences serve to increase or decrease the overall rate of bone recycling, while other influences stimulate or inhibit only one arm—osteoblasts or osteoclasts—of bone recycling.

The lifetime cycle of bone growth and deterioration varies by gender, race, diet, exercise, disease, tobacco use, steroid use, and genetic predilection, but the overall pattern is universal: increasing

CLINICAL CHANGES IN BONE

In osteoporotic bone, we see three changes, all of which result in poor bone quality and decreasing strength:

1. The cortex (the thick outer layer) of the bone becomes thinner.
2. The cortex also becomes more porous.
3. The medullary (inside) portion of the bone becomes more porous and less well-connected, having less and less of the trabecular matrix.

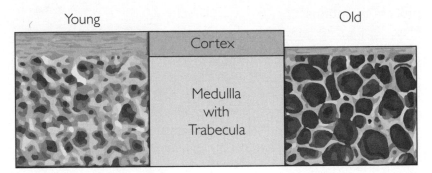

Bone changes with age: osteoporosis.

mass in the young, maintenance of bone mass in the adult, and then a gradual loss of bone mass, which we characterize as osteoporosis, in the elderly. Overall, however, the bone loss is not a result of endocrine changes, but of changes related to the aging process itself, at the cellular level. The onset of bone loss—with the imbalance in favor of osteoclasts over osteoblasts—begins prior to the decline in sex steroids (such as estrogen and testosterone) and is accompanied by a parallel slowing in bone turnover as a whole. In women, once menopause occurs, both changes become even more significant; in men, there is a more gradual loss of bone and bone turnover, given the more linear nature of "andropause."

OSTEOPOROSIS: QUICK FACTS

Age: Generally begins after age forty or after menopause in women. May be present in 50 percent of those over age fifty, and is overwhelmingly common (though more so in women) by age seventy-five. Spontaneous and traumatic fractures become increasingly common with age, and many older people show a "dowager's hump" as their upper vertebrae fail.

Costs: US annual costs are estimated at more than $22 billion, including joint replacements, although costs for such procedures are hard

continued on next page

to separate from those due to osteoarthritis. US and global costs are rising steadily as people live longer.

Diagnosis: Most patients are diagnosed after incurring an unexpected fracture, typically a vertebra, wrist, or hip. Often, risk can be assessed simply by clinical history and known risk factors. Diagnosis can be confirmed using standard X-rays or by measuring the bone mineral density (BMD) using a bone scan. The disease is usually diagnosed when the bone density is several (often 2.5) standard deviations below normal.

Treatment: While the risk of osteoporosis—and resultant fractures—can be lowered by having an active lifestyle and adequate diet, avoiding steroids and tobacco, and using bisphosphonates, most of these only slow progression at best. There is no current intervention that can reverse or even stop the progression of osteoporosis, although treating cellular aging at the genetic level holds promise.

Bone loss—osteoporosis—is not simply a passive event that occurs with age; it is a disease. As such, it causes a gradually increasing risk of fracture with advancing age, so much so that if we were to live long enough, that risk would reach one hundred percent as our bones were gradually lost completely.

Muscle Aging

Muscles lose both mass and strength with age—a statement that is both true and enormously oversimplified. Muscle aging is a very complex process involving muscle tissue and other systems.

For example, the aging blood supply to muscles—and the resultant aging of muscles—can cause unexpected pathology in other systems as well. Even if muscles did not age independently, which they do, they would show gradual failure with advancing age as the blood supply, endocrine system, nervous system, joints, and bones began to fail. We have already discussed osteoarthritis and osteoporosis, which have implications for the mechanics of our musculature, but the most prominent effect on muscle aging occurs as the vascular system ages, resulting in less reliable access to oxygen, sugar, and other nutrients needed for muscle activity, as well as slowing the removal of carbon dioxide and other waste products. Relative denervation of the muscle can also play a role as the peripheral nervous system "prunes" some of the efferent connections (efferent nerves carry impulses *from* the brain), thereby making our movements less precise and coordinated.

The aging of muscles also degrades the function of other systems. As the muscles age, they use less energy; as the overall energy expenditure decreases, predilection increases for obesity (especially abdominal fat), which increases insulin resistance and the risk of type 2 diabetes, as well as hypertension and cardiovascular disease. In addition, the muscles represent a large storage depot of available protein for the rest of the body. As muscle mass declines with age, such proteins are less available to meet the emergent needs of the immune system (for enzymes and antibodies), the liver, and other organ systems, and this loss of mass is a predictive factor for mortality in the elderly.

Also, the aging process is complex within the muscles themselves. The most obvious effect is simply the loss of muscle mass, due to inadequate replacement of damaged fibers and to shrinkage of the remaining fibers. Similar to what we have seen in other systems, the muscle fibers of young adults are replaced as fast as they are lost, but as we age the rate of replacement fails to keep up. Also, the replacement tissue is often fat or tough, fibrous tissue rather than actual muscle. The outcomes are smaller muscles and reduced strength, even if muscle mass *does* remain the same.

While these changes are obvious by exam or physical testing, there are many problems occurring within the aging muscle that are more subtle, but which underlie the more obvious problems. In addition to the decrease in the quantity of muscle, there is a definite decrease in its quality. This is most prominently seen in decreases in protein synthesis and oxidative capacity.

Protein synthesis decreases with age in almost all cells, although the changes in rates of synthesis vary between proteins. The effects are—as we have already stressed in previous chapters—manifold and often unexpected. The obvious effect is that repair is slowed.

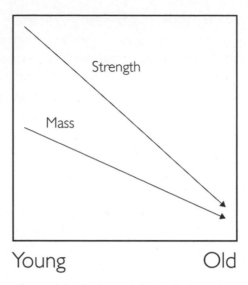

Young   Old

As we age, muscle mass and strength both decline, but strength declines more rapidly.

The less obvious is a gradual decrease in the quality of the available proteins, resulting in dysfunctional cells, which causes a loss of strength and contributes to the decrease in oxidative capacity.

The overall metabolism of the aging muscle shows a measurable decline, and this is particularly evident in mitochondrial function. The mitochondria—the key to cellular energy—become fewer, and each of them becomes less effective. The overall decline in available energy—especially ATP—results from decreases in the protein turnover within the mitochondria, because most of the mitochondrial proteins depend on gene expression within the nucleus, and gene expression has slowed with age. Mitochondria become less and less capable of producing energy for muscles. Accompanying these changes, the mitochondria show decreases in their oxygen uptake and in the activity of the enzymes responsible for oxidation. This is predictable; as protein turnover slows, the available proteins are more likely to be damaged proteins. As a result of having less available ATP, older muscles have less endurance and less strength, but this may play an additional role by limiting the energy available to produce proteins in the muscle, further limiting protein turnover and cellular repair.

The most marked change in protein turnover is among the relatively uncommon proteins that are critical to energy metabolism, but even the more stable proteins have a slight decrease in turnover, which also causes a loss of strength. Myosin (also myosin heavy chain, or MHC), a key protein involved in muscle movement, has a low rate of turnover, but it becomes even lower in the aging muscle with the same result: an increasing percentage of incompetent proteins with a resulting loss of muscle quality. In both the young and the old muscle, the rates of turnover respond to exercise, particularly aerobic exercise. While resistance training can increase muscle mass, aerobic exercise improves protein turnover and therefore muscle quality rather than quantity. In any case, these benefits become smaller with age: For a given amount of exercise (aerobic or otherwise), young muscle will generally show greater benefit than will old muscle. And even given a steady level of exercise, muscle mass and strength will decline with age.

There has long been a belief that much of muscle aging could be prevented or reversed by exercise. This is true only to a limited extent. A sedentary middle-aged or elderly person will certainly increase muscle mass and strength through exercise, but this becomes increasingly difficult and the benefits increasingly limited as a person ages. In short, there is always some benefit from exercise. Without exercise age-associated loss of mass and strength is worse, *but exercise does not prevent or reverse muscle aging itself.* In other words, even though exercise has no effect on actual muscle aging, exercise helps elderly patients function better.

But maybe not all of them. There is a large subset of the elderly who simply cannot grow muscle, apparently due to age-related changes, and this subset increases with age as well. Oddly enough, this is likely due to an interesting feature of muscle: the fact that it derives from a subset of muscle stem cells, or myocytes. It was once universally understood that muscle cells, like nerve cells, don't divide after birth. Then we discovered that there are exceptions, that we do see some cases of muscle and neuron cell division even in adults. But the question remained: How important was such cell division in practical terms? You might reasonably think of this as a mere academic issue, but it turns out to have important clinical implications. Telomere shortening—and hence cell aging—occurs almost exclusively among cells that are dividing, so the obvious question was whether or not muscles "really aged." That is, did their telomeres shorten as we grew older?

Existing muscle cells, myocytes, derive originally from myoblasts (a type of muscle stem cell) or satellite cells (a type of muscle stem cell found in the adult muscle tissue), so the muscle cells in the elderly have certainly undergone cell division and hence telomere shortening. Actually, there is evidence that a more generalized, pluripotent type of stem cell may circulate in the body and that some of these can not only differentiate into muscle stem cells, but can become perfectly normal and functional muscle cells, including cardiac muscle cells (cardiomyocytes).

MUSCLE AGING: QUICK FACTS

Age: Muscle loss often begins in early adulthood, probably as a result of decreases in physical activity, and becomes more marked in women after menopause. Even in otherwise healthy people, the loss of both muscle mass and the number of muscle fibers is apparent by age forty if not earlier. The process is gradual and progressive, without the obvious inflection points that are seen in some other age-related diseases, such as fractures, heart attacks, etc.

Statistics: Costs and other data are difficult to obtain, because they derive mostly from increased frailty and are attributed to other medical problems: falls, fractures, joint replacement, secondary infections. Also, loss of muscle probably leads to increases in diabetes and other illnesses.

Diagnosis: Muscle aging is most noticeable in the hands, which may appear thin and bony.

Treatment: "Use it or lose it" tends to be the recommended therapy, although the potential benefits of exercise vary markedly between individuals, and as we age, it becomes easier to lose muscle through inactivity and harder to gain it through activity. There is no other effective therapy.

So the gradual loss of muscle quality and quantity that occurs in the elderly—in both skeletal and cardiac muscles—reflects a gradual exhaustion of our ability to use stem cells from several sources to replace muscle. Muscle aging results from much the same process as occurs in many other systems—bones, joints, skin, and so forth—as the telomeres shorten and induce a gradual and currently irremediable loss of function.

Skin Aging

There is a common, and somewhat inaccurate, perception that skin aging is merely cosmetic. This perception derives from two sources. The first is that while we are well aware of friends and relatives dying of heart attacks, strokes, cancer, and Alzheimer's disease, we are usually unaware of anyone who has died of "old skin." The second is that we are barraged with advertising for various creams, lotions, drugs, and treatments that purport to "erase wrinkles," "make your skin young again," or "fight both visible and future signs of aging." People spend billions of dollars on these products. Some of them, Botox, for example, do have measurable cosmetic effects, but many of the most widely used products lack any basis for their claims, yet sell well, and at extravagant prices. Even when "anti-aging" skin products like Botox work as advertised, the results are merely cosmetic, and our concern here is the medical aspects of aging skin.

First, it might surprise you to know that people do occasionally die of aging skin. Extreme skin aging can result in loss of a barrier to infection, so that patients die of infected skin lesions. While even the young can die of infections that penetrate the skin barriers, this cause of death increases with age, as the skin not only becomes a less secure physical barrier, but no longer has the support of an adequate blood supply and an effective immune response.

More typically, however, aging skin is not so much the cause of death as it is an important contributor, for several reasons. Aged skin is not as effective as a physical barrier or as a thermal insulator, so the body wastes more energy to maintain a normal temperature. Its abilities to heal injuries, to sense injury, and to respond with an immune defense are all diminished. These and other changes in aging skin result in a rapidly increasing burden on the remainder of the body, further stressing other systems that are themselves aging and losing competence.

There are two types of cells that fundamentally define skin: fibrocytes and keratinocytes. There are actually dozens of other cell types that are found in normal skin, including cells for specialized structures such as hair follicles and sebum glands. There are cells that enter the skin from elsewhere, such as blood vessels and nerves, and transient cells (also called "wandering" cells) that arrive from the bloodstream. Keratinocytes make up the outer layer of the skin—the epidermis—and are continually dividing, replacing lost cells, and being sloughed off in our day-to-day lives. Therefore, their telomeres shorten continually over our lifespans, with consequent age-related changes in the epidermis.

The inner layer—the dermis—is more complex and includes fixed cells, such as fibroblasts, as well as wandering cells— macrophages, monocytes, lymphocytes, plasma cells, eosinophils, and mast cells—that generally serve immune functions. The fibrocyte is the key fixed cell within the dermis. These cells can divide and form fibroblasts and fat cells. The fibroblasts create and maintain the extracellular matrix of collagen and elastin fibers that holds the layer together. Fat cells—adipocytes—are more common in young skin and protect the body both as a physical cushion and as a thermal insulator. Both fibroblasts and adipocytes result from cell divisions of the fibroblasts. As these cells are lost, the fibrocytes divide and replace them with newer fibroblasts and adipocytes, but with a gradual loss of telomere length and—once again— age-related changes in the dermal tissue.

Our skin cells—in both the dermis and the epidermis—change their pattern of gene expression as we get older, becoming slower to divide, fewer, and less capable of performing their roles. The fibroblasts, for example, become slower at their job of replacing the extracellular matrix of collagen and elastin fibers and more prone to produce altered fibers. The result is that skin loses both its strength (the collagen) and its elasticity (its elastin) as we age. Likewise the adipocytes become fewer, causing an overall loss of skin fat, while the keratinocytes divide more slowly, failing to offset the rate of cell loss—causing a decreasing population of cells in the epidermis.

These changes are obvious just in the way skin looks and feels as we age. We find that older skin is slower to heal, as the cells no longer divide as quickly. It tears easily, as the collagen fibers no longer provide the strength they had in young skin. When we pull or lift the skin, the elastin fibers no longer pull it back into place as quickly (or at all). We begin to see "bags" and "tenting." As we lose fat cells in the dermis, the skin—now lacking an adequate cushion—is more prone to injury, even with minor bumps or scrapes, and bruises more easily. The loss of fat allows the body to lose heat more quickly, so we become chilled easily, which stresses our metabolism by requiring more calories to maintain a normal body temperature.

One of the most common changes, however, is in neither the epidermis nor the dermis, but in the boundary between the two layers. In young skin, the dermal-epidermal junction is interdigitated, which means the two layers intertwine like fingers, resulting in a strong mechanical junction that makes it all but impossible to pull the epidermis away from the dermis. Young skin is tough. In old skin, however, this interdigitation is progressively lost, and the boundary—instead of being intertwined—becomes almost flat, with pockets of fluid (microbullae) where there is no adhesion at all. The result is that the skin of the elderly can easily slough away with only the slightest friction. If an elderly person trips and falls, for example, she may pull away large sheets of "tissue paper skin" as her forearm slides across the edge of a chair on the way down. Old skin is notoriously delicate, easily torn, and almost impossible to suture back into place.

Aging skin becomes thin, weak, and damaged over time, particularly in areas exposed to the sun. We see wrinkles and age spots, or liver spots, in which the skin has lost effective control of its pigmented cells, resulting in small, irregular areas of darker discoloration. The cause of wrinkles and "dry skin" is not a loss of moisture, but a loss of skin cells and damage to the extracellular matrix. As skin cells no longer repair and replace damage, areas of frequent micro-trauma, where the muscles repeatedly flex the skin, as on the face in areas used to show expressions, begin to display prominent and lasting changes that we see as wrinkles. The same effect—but more scattered and finer—occurs throughout the rest of the skin, as on the backs of the hands and the forearms, which show thousands of tiny parallel wrinkles as the skin loses elasticity and cellular volume.

AGING SKIN: QUICK FACTS

Age: Skin aging is a cumulative, lifelong process, with no specific age of onset. Rate of aging varies between individuals, but other than age itself, the cause is exposure, particularly to ultraviolet radiation, which accelerates telomere shortening and skin aging.

Costs: Primary medical costs are related to increased rate of injury, infection, and metabolic stress, as well as decubitus ulcers in those confined to bed, but it is difficult to separate these costs and assign them specifically to aging skin. On the other hand, Americans spend $20 billion a year or more on cosmetics, drugs (such as Botox), and plastic surgery in hopes of making the skin appear younger.

Diagnosis: We hardly need a doctor to tell us when our skin has aged. It's the first thing we look at in evaluating age in ourselves and others. Medical evaluation of skin aging is generally done by measuring skin elasticity, which is a good indicator of aging in general, but can also be evaluated by biopsy. Also, skin aging can result in ulcerations—particularly in bedridden patients—commonly evaluated by depth, size, and penetration into deeper structures, as well as for active infection.

Treatment: All the procedures and products used to combat skin aging have at best only cosmetic effect, with only a couple of possible exceptions. One is products containing retinoic acid, which increase the rate of cell turnover (and may actually hasten telomere shortening). The other is TA-65 (discussed in Chapter Four), which, among those products claiming to slow telomere loss or even re-extend telomeres in skin cells, is the only one shown to have any effect. UV protection (sunscreen) does slow skin aging by reducing UV damage to the skin.

Endocrine Aging

Lowenstein! The name brought back to me the memory of some snippet from a newspaper which spoke of an obscure scientist who was striving in some unknown way for the secret of rejuvenescence and the elixir of life. . . . When I have written to this man and told him that I hold him criminally responsible for the poisons which he circulates, we will have no more trouble. But it may recur.

—Sir Arthur Conan Doyle,
"The Adventure of the Creeping Man"

History is replete with claims that various hormones will reverse aging. The majority of these claims, particularly early ones, focused on sex hormones, usually hormones derived from male gonads. In Sir Arthur Conan Doyle's "The Adventure of the Creeping Man," a professor uses an extract from langur monkeys to regain his youth and sexual ability. Doyle's fictional story has its roots in the work of the French surgeon Serge Voronoff, who was infamous for grafting monkey testicles into his patients a century ago, after failing to get results with hormonal extracts alone. Voronoff even published a book on this work; *Rejuvenation by Grafting* is long out of print, while Doyle's Sherlock Holmes stories live on and are doubtless better reading.

Voronoff's wasn't the last such folly. Now, instead of monkey glands, it is the supposed benefits of growth hormone, melatonin, and androgens, and the public seems equally credulous despite a lack of data showing any actual effects on aging.

For many years, the endocrine system has been alleged to be the cause of aging and hormone replacements hyped as a cure. However, there is no evidence that hormones cause aging, much less that they can cure it or even affect it in any way. Certainly, hormones can be beneficial in the right situations and in appropriate doses, but hormones also can cause disease or death in the wrong situations or in inappropriate doses. Perhaps the most common endocrine approach—though it is still in dispute and carries obvious risks—is hormone replacement therapy (HRT). HRT is often assumed to refer solely to the use of estrogen or progesterone in women after menopause. This indication was first suggested just prior to the middle of the twentieth century, as pharmaceutical sources were developed, and its use had become endemic in the United States by the century's end. Many patients felt that HRT postponed aging and age-related diseases, although the preponderance of data suggests that this is illusory. Worse yet, some trials suggest that the use of HRT in women increases the risks of Alzheimer's disease, breast cancer,[1] stroke,[2] and heart disease, although it may depend upon when HRT is initiated.[3]

Use of other hormones to stave off aging began just after the mid-twentieth century, particularly growth hormone, then accelerated in the 1980s and 90s as it began to be produced commercially. Melatonin has had a commercial run as the "fountain of youth" despite a complete absence of supportive data for any such benefit, not to mention the fact that melatonin levels probably don't change with age in the first place. Now hormone replacement "therapy" has become the presumed fountain of youth in the minds of people as gullible as those who turned to monkey glands a century ago. Despite public credulity and the very profitable commercial market for "anti-aging hormones," there is no evidence that any hormones have any effect on aging per se. Moreover, there is no logical argument to suggest that hormones *should* affect aging. As discussed in Chapter Three, if an endocrine gland were responsible for timing the aging process, it would merely beg the obvious question of what timed aging inside the endocrine gland in the first place? These odd public fashions—based almost entirely on wishful thinking—have been with us in the past, are with us currently, and will be with us in the future, despite the lack of any evidence or even rational speculation.

---

[1] http://www.breastcancer.org/risk/factors/hrt
[2] http://onlinelibrary.wiley.com/enhanced/doi/10.1002/14651858.CD002229.pub4
[3] http://link.springer.com/chapter/10.1007/978-3-319-09662-9_18

On the other hand, while the endocrine system doesn't orchestrate aging, many of the hormones in our bodies show a pattern of change as we get older. Some hormones show a slow decrease (as with testosterone levels), some show a sudden decrease (as with estrogen levels at menopause), and many show an increasingly aberrant pattern of response to physiological stimuli (as with epinephrine, thyroid, and some other hormones). While most advocates of hormone replacement know that many hormones show a decrement with age, very few appear to be aware of the increasing and probably more significant loss of physiological control that occurs in hormonal responses as we age. The key to the dysfunction is not that we have lower levels of hormones circulating in our bloodstreams, but that those hormones are no longer responding as rapidly or appropriately as they should.

Moreover, many of the most important endocrine problems that occur with age go almost unnoticed by proponents of wholesale HRT. The increasing problems with insulin resistance, for example, occur in the cells, not in the bloodstream. The all-too-common problem of type 2 diabetes (a hallmark of aging in many patients) is associated with changes in insulin levels (and the body's inability to adjust insulin levels rapidly and accurately in response to changes in blood sugar). However, much of the problem lies in the ability of aging cells to respond to insulin levels. Similar problems occur with other endocrine systems, in that the most notable problem we face with aging is not changing blood levels of the hormones, but a changing (and deficient) cellular response to those levels. These more complicated changes probably play an important role in many other metabolic dysfunctions as we age, as with "metabolic syndrome," hypercholesterolemia, increased LDL, calcitonin[4] responses, and the changes underlying many cases of obesity.

Finally, there is a common assumption that if any hormone level goes down with age, then increasing the level to that found in a young person will obviously make you healthier. This assumption is so common that it is almost never questioned by HRT proponents;

---

[4] A hormone central to bone maintenance and osteoporosis.

their only question is how much hormone to administer to "get you back to normal." While this reasoning may be true where hormone changes are the actual *cause* of a disease—as in primary growth hormone deficiency, in which a child simply fails to make enough growth hormone for normal development—there is no reason to make the same leap of faith where the hormonal change is secondary and not the cause. To cite a simple example, in the case of type 1 diabetes, loss of ability to produce insulin is the cause of the disease, and raising blood insulin levels *appropriately* can prevent death by hyperglycemia (although it doesn't necessarily prevent long-term health problems). The insulin level of a person with acute diabetes is chronically low and needs replacement. On the other hand, fasting also lowers insulin, yet giving insulin to a fasting person can be fatal. In this latter case, low insulin levels are not the primary problem, but merely a secondary result of having low blood glucose from fasting.

The same general argument can be made regarding many of the changes that occur with aging. Proponents of HRT would argue that if your testosterone or estrogen level falls with age, it would be beneficial to raise it. However, if such age-related changes are secondary, then we would expect no improvement. And if such changes are protective, we could actually cause significant health problems. The problem with HRT is not that it might not be beneficial in some cases, but that it's usually based on the weak assumption that low hormone levels are bad. In assessing the potential benefits of HRT, we should be looking at data, not thoughtless assumptions. As noted earlier, the current data suggest that raising estrogen levels to the premenopausal "normal" results in a much-elevated risk of Alzheimer's and other age-related diseases. Normal must be defined in relation to age and in terms of the outcome. Estrogen levels are certainly not "normal" if they cause disease and result in a higher mortality rate.

In case I haven't made this clear enough, I'll say it one last time. Specific hormones may offer specific benefits for specific problems, but *hormone replacement has no effect on aging.*

ENDOCRINE AGING: QUICK FACTS

Age: The only unarguable age-related changes in hormone levels are the inflection that occurs with menopause and the lesser and more gradual decline in androgenic hormones. There is a decline in growth hormone levels in most people, although whether it is related to aging, inactivity, changing sleep patterns, diet, or disease is in dispute.

Costs: Various estimates suggest that the global market for anti-aging endocrine therapies is now approaching $3 billion per year.

Diagnosis: Accurate diagnoses of hormone deficiency can be made on the basis of laboratory testing, although in some cases this requires special preparation or recurrent testing.

Treatment: For most hormones, treatments are available for cases where the level is lower than age-normal, but there remains substantial disagreement on the advisability of such therapies and a growing consensus that the risks usually outweigh the presumed benefits. *Caveat emptor.*

Localized Systems and Special Cases

Pulmonary Aging

The older we get, the harder it is to breathe. Many older people are unaware of any significant change in their breathing capacity, because at rest or during most common daily activities, we use only a small part of our pulmonary capacity. When we exert ourselves, however, we expose any concealed pulmonary problems we might have. Aging first cuts into our residual lung capacity, and only later affects breathing in our routine daily activities.

Our lungs age slowly and progressively—and independently of smoking, injury, infection, or other pulmonary insults. In youth, we notice shortness of breath only during extreme or prolonged exertion. With age shortness of breath occurs with progressively lower levels of exercise. Also, any history of pulmonary problems rapidly increases this process, with the result that many chronic smokers become short of breath with no exertion at all.

Aging itself—again, independent of other problems—results in several types of changes within our lungs. Prominent among these are structural changes, vascular changes, and immune changes. Vascular and immune system aging are discussed elsewhere in this book, so we'll look at structural changes now.

Age-related structural changes in the lungs are almost completely defined by a progressive loss of alveolar surface (as in COPD), although changes also occur between the alveoli as well (interstitial lung diseases). The alveoli—the small sacs that permit gas exchange with our blood supply—become fewer with age. Imagine two small soap bubbles, joining to form one slightly larger soap bubble. Much the same occurs with many of our tiny alveoli; the result is a smaller area for exchange of gases and far less effective lungs. A large part of this problem results simply from the loss of alveoli themselves, but the lung tissue also loses elasticity, support, and muscle function, resulting in narrowing of the small airways. These two problems cause small airways to close off, further decreasing available alveolar surface. The cumulative result of these factors is that while the total lung volume may remain fairly constant with age, the number, surface area, and complexity of the alveoli *within* the lung all decrease, so that it becomes harder and harder to maintain effective gas exchange. It requires more and more exertion to keep our bodies oxygenated and to remove carbon dioxide from our bloodstreams. Carbon dioxide levels in the blood may slowly increase, while oxygen levels may slowly decrease.

The resultant symptoms—especially the subjective shortness of breath—are perhaps the most terrifying of all age-related problems. Shortness of breath, like drowning or suffocation, strikes to the heart of our deepest fears, engendering panic. Acute strokes and heart attacks bring sudden death. Alzheimer's disease is tragic, and other age-related diseases are disabling, but shortness of breath is acutely terrifying, and that dread is constantly present as our lungs fail. Fortunately, this degree of pulmonary failure is uncommon during normal aging. Most of those who have significant symptoms have been smokers or have had other severe pulmonary problems in the past. Nonetheless, if we lived long enough, if we didn't succumb to other age-related causes of death first, we would all share these symptoms. The only way to prevent them is to circumvent the basic cellular causes of pulmonary aging.

As we age, not only do we have fewer cells in our lungs, but those cells that are still present show shorter and shorter telomere lengths. This is true of the cells that make up the alveoli themselves as well as other cells within the lungs, such as interstitial cells, immune-related cells (macrophages, for example), and cells that make up the capillary walls. In all cases, these effects—including the telomere shortening—are accelerated by smoking, severe and recurrent pneumonias, and other pulmonary insults.

The most commonly diagnosed age-related pulmonary disease is usually called COPD (chronic obstructive pulmonary disease), although the terminology has changed over the years. This diagnosis often overlaps with emphysema, idiopathic pulmonary fibrosis, diffuse interstitial fibrosis, interstitial pneumonia, and others. The confusion and diagnostic overlap lie in the fact that age-related pulmonary changes are actually a spectrum of changes from the alveoli on the one hand (as in COPD) to the tissue between the alveoli on the other (interstitial lung diseases). While all such changes result in decreased pulmonary function, they may have different patterns of presentation, different patterns in their diagnostic hallmarks, and somewhat different courses. In almost all cases, however, this disparate collection of diseases has a common denominator: They are age-related and exacerbated by any pulmonary insult such as smoking, pollution, infections, etc. Most important, for the person with the disease, the outcomes are the same—failing pulmonary function, shortness of breath, inability to perform activities of daily life, and a high risk of mortality.

Age-related pulmonary diseases likely share the same cellular pathology: damage, loss of cells, shortening of the telomeres, changing patterns of gene expression, altered cell function, impaired tissue function, and consequent overt clinical disease. A history of pulmonary insult—tobacco use, pollution, infections, etc.—results in a more accelerated pattern of change in pulmonary cells. These insults damage and kill cells within the lungs, forcing a more rapid rate of cell division for replacement, which accelerates telomere shortening, epigenetic changes, and the onset and severity of age-related pulmonary diseases. Other than symptomatic

PULMONARY AGING: QUICK FACTS

Age: Beginning in our twenties and thirties, there is a measurable loss of alveoli. The subsequent rate of loss is greater in men than in women. Interstitial changes—in the tissues between the alveoli—similarly change with age, but are often diagnosed later and may have a more rapid course.

Statistics:[1] COPD is diagnosed in about 5 percent of the population. In some developed countries, it is the fourth leading cause of death. The disease will likely increase due to increases in tobacco use in developing countries and globally as more people live longer.

Costs:[2] Estimate for the US alone is approximately $50 billion per year.

Diagnosis: Diagnoses are often made on the basis of symptoms (such as shortness of breath, cough, and sputum) and physical examination, and are generally supported by X-rays, pulmonary function tests, arterial blood gases, and high-resolution computed tomography (HRCT), especially in the case of interstitial lung disease. In some cases, pulmonary biopsy is used.

Treatment: Current treatments are largely supportive. While these treatments may offer acute symptomatic relief, they have little or no effect on the overall progression of the disease. In all cases, patients should quit smoking and avoid other factors that accelerate pulmonary damage, such as air pollution and infection. Treatment may include antibiotics (acute bacterial infection), vasodilators, steroids, vaccinations (to prevent pneumococcal or viral infection), oxygen therapy, pulmonary rehabilitation, or—in some extreme cases—lung transplantation.

---

[1] http://www.cdc.gov/copd/data.htm
[2] http://www.lung.org/lung-disease/copd/resources/facts-figures/COPD-Fact-Sheet.html

therapy, there is no current clinical intervention that prevents, halts, or significantly slows age-related pulmonary disease.

As with other systems, an effective therapy will require the ability to re-extend the telomeres of our pulmonary cells.

Gastrointestinal Aging

The gastrointestinal tract extends from the mouth to the anus, a collection of disparate tissues with nutrition-related functions: to bring in food, break it down, absorb the nutrients, and eliminate the leftovers. Topologically, the human body is a doughnut—a toroid—with the GI system representing the hole—or tube—running through the center of the doughnut, albeit a very complex tube in both its shape and its functions.

At the mouth, age-related changes are typically dental, including an increasing incidence of periodontal disease and gingivitis. The temptation is to simply ascribe all age-related dental changes, including gradual erosion of the enamel and loss of teeth, to "inevitable wear and tear." To a large extent, this is justified and—beyond the initial replacement of childhood teeth by the adult set—there is little reason to think that much can be done to halt or reverse the problems caused by use alone. On the other hand, many age-related dental changes are magnified by dietary risks (sugars, acids, etc.) and a lack of brushing and flossing. There is good reason to think that immune-system aging plays a role as well, certainly in periodontal disease where chronic low-level infections are key players in the pathology. Data shows that telomere shortening correlate with both periodontal disease and immune aging.

Whether as a result of genetics, diet, or hygiene, some people manage to keep their teeth more or less intact to an advanced age, while others lose most or all of their teeth early in adulthood, even before other systems demonstrate age-related changes. While it is certainly possible that re-extending telomeres in immune-related cells would markedly improve oral health—specifically, prevent periodontal disease—improving the retention of healthy teeth into advanced age, the evidence also implies that both diet and oral hygiene will always remain major predictors of age-related dental problems.

In the liver and intestines, age-related changes are often difficult to separate from diseases not related to aging, because many of these diseases begin or become worse as we age, including gastroesophageal reflux and various bowel diseases such as Crohn's disease, regional enteritis, irritable bowel disease, etc. In many cases, these diseases may be triggered or exacerbated by cell aging within the gastrointestinal system or in the immune system, but there is no obvious reason to discuss most of these diseases within an aging context.

There are, however, age-related changes that occur in the intestines. Most of these changes have to do with the functions of the intestinal walls themselves, rather than with the diseases mentioned above. Aging intestines—even in the absence of a specific disease—show significant losses in absorption, immune function, and motility, for example. The older intestine is less capable of absorbing nutrients and less capable of producing the various enzymes and co-factors required to perform such absorption effectively, particularly for iron, calcium, zinc, and vitamins B_{12} and D. Oral medications may also be absorbed poorly, making drug levels unreliable or insufficient. The bowel walls lose muscle strength, making peristalsis—the waves of contraction that move food down the intestines—less effective and increasing the risk of constipation. Bowel walls also lose strength and elasticity, making it more likely that peristaltic waves will cause the bowel walls to distend outward in small balloon-like pouches that expand out through the normal wall. These diverticula become inflamed (diverticulosis) or infected (diverticulitis) and can cause significant morbidity or mortality in the elderly patient. More than half of people over seventy[5] already have diverticulosis, and this often progresses to complications.

In general, the cells of the gastrointestinal tract—specifically the cells responsible for absorption and production of co-factors, the muscles, and the immune cells—divide and show telomere shortening. Re-extension of their telomeres can be expected to improve age-related changes.

---

[5] http://www.lung.org/lung-disease/copd/resources/facts-figures/COPD-Fact-Sheet.html

Urogenital Aging

The kidneys, bladder, and associated structures show important age-related changes. Some of those changes are most important to you personally, such as your kidney function, while others affect people who live with you—such as waking the person with whom you share a bed when you get up to go to the bathroom several times a night. And of course, the bladder isn't the only urogenital structure that tends to fail with age, with implications for both you and the person you share your life with.

The job of the kidneys is to filter the blood, put back what you still need and excrete what you don't. Both of these missions— taking things out and putting them back—become less effective as you get older. As we lose cells and replace them, our remaining renal cells have shorter and shorter telomeres. These older cells are both fewer and less effective. Loss of kidney cells results in a loss of nephrons, the filtering units of your kidneys that do the actual work. As in other organs, the arterial walls and the capillary beds also show age-related changes. Taken together, the changes in the kidneys and the blood vessels increase the risk of hypertension as we age. With fewer nephrons and aging cells, kidney efficiency goes down. With aging arteries and ineffective filtering, blood pressure goes up. Finally, the levels of several important molecules in our blood that have to be kept within careful limits by the kidneys tend to go awry. Even when these levels are normal, they tend to be less stable and to become abnormal far more easily than in young adults. In most people, the old kidney has enough reserve to manage day-to-day needs, but renal failure becomes more likely as the kidneys lose their reserve function. The older we are, the less stress it takes to cause a significant renal problem, including renal failure.

Not only does the bladder itself lose cells and cellular function, but the cells of the bladder wall become less capable of maintaining elastin and collagen. As a result, the bladder becomes less elastic, less distensible, and less able to hold as much urine as when we are young. The muscles become weaker and less able to completely and rapidly empty the bladder. As a result, we are increasingly unable to get through the night without needing to urinate. All of these changes, along with the aging of the immune system, make us more and more likely to have urinary infections, incontinence, or retention.

In both sexes, dramatic changes in the aging urogenital system have a marked impact on the ability to support sexual intercourse. In men, the ability to begin and to sustain an erection becomes impaired as they get older. While a number of factors can accelerate the onset and severity of erectile dysfunction—such as obesity, smoking, alcohol use, and insufficient exercise—there is little doubt that aging itself is responsible for much of the problem. Once again, the cells responsible for maintaining the vascular changes necessary for an erection lose their function as cells divide, lose telomeres, and undergo a changing epigenetic pattern. In women, the most marked change in the vaginal mucosa occurs in conjunction with menopause, but even here, the mucosal cells have divided and lost telomere length. The changes in the epigenetic pattern are in this case the result of two factors: shorter telomeres and the lower levels of estrogens. Estrogens, like other steroid-based hormones, directly bind to the chromosomes and modulate gene expression. As a result, the vaginal mucosa are thinner, with less muscle and elastic function, as well as less able to produce lubrication.

Sensory Aging

The changes in the sensory system are legion, including touch, vision, hearing, smell, and taste. Changes in touch sensation are often unnoticed—probably because they are very gradual and play a lesser role in daily activities and social interactions. The decline of smell and taste is also gradual, but it is more noticeable, especially as we mourn the loss of enjoying food, for example. Worst of all, though, is the loss of acuity in our vision and hearing. Sight and sound are central to most of what we do every day of our lives in work, play, and social interactions. The loss of vision and hearing as we age is keenly felt.

We note the loss of overall ability in any sensory system, but that loss is seldom one of sensitivity *within* the receptor, but rather in the ability to discern *between* receptors. In the case of touch, for example, the individual sensor may be just as sensitive, but the number of receptors is decreased. This principle is almost uniform and often misunderstood. For example, there are two ways in which we can lose hearing. Loss of receptor sensitivity makes it hard to hear soft sounds. Lost ability to discern between receptors makes it hard to tell words apart when we listen to speech.

Touch

In the case of touch, we lose receptors as we age. We are born with a set number of receptors, which means that the number of receptors per square millimeter of skin decreases as we grow up and our total skin area increases. Aging, however, creates a more important change in that we actually lose receptors. Any given receptor might be as sensitive as ever—to light touch, for example—but our ability to precisely locate the point of contact is diminished by the loss of receptors. We still know we have been touched, but aren't as sure of precisely where or by what. This is most precisely measured by "two-point discrimination"—the ability to tell the difference between a single touch and two simultaneous touches that are just a small distance apart. Subjectively, however, the loss is more noticeable in our ability to identify, say, types of fabric or objects in our pockets or purses. In short, we become gradually less adept at identifying objects by touch. Data suggests that, as a result of the more than 80 percent decrease in the number of touch receptors per square millimeter of skin, we are only half as accurate at identifying objects by touch at age seventy as when we were young.[6] If we look closely within the skin, we find that the number of nerves present is often only marginally changed, while the number of receptors per nerve and the rate of conduction are both markedly decreased with age. In short, as we age, we are slow to notice and are poor at telling what we have touched compared to when we were younger.

---

[6] http://informahealthcare.com/doi/abs/10.1080/08990220310001622997
http://informahealthcare.com/doi/abs/10.1080/08990220601093460
http://www.scholarpedia.org/article/Touch_in_aging#Changes_in_touch_sensitivity_and_spatial_resolution_with_age

Both these changes are likely attributable to cell aging. While the peripheral nerves rarely divide and are therefore unlikely to show cell aging, the cells that create the myelin sheaths around our peripheral nerves, ensuring rapid transmission, do show cell aging, as do the cells responsible for the peripheral touch receptors themselves. Such peripheral receptors—whether for light touch, pain, temperature, or pressure—are all subject to replacement during normal use and therefore subject to cell aging as their telomeres shorten.

Smell and taste

As we age, we gradually lose the ability to discriminate odors and flavors. As with the peripheral sense of touch, the loss is secondary to the aging of the receptor cells—the olfactory receptors in our noses and the taste buds in our mouths. While there is little loss of our ability to sense pungent stimuli—bitter flavors and unpleasant smells—we are far less capable of discerning subtle differences of flavor or smell, and this loss is particularly notable in the common daily experience of eating. Foods become less enticing and less enjoyable.

Olfactory ability clearly declines with age, but precise measurements are difficult to make; it's hard to quantify something as subjective as smell. Nonetheless, most studies show that a loss of olfactory discrimination definitely occurs in many people by age seventy and becomes common and severe in those above age eighty. This change corresponds to a notable decrease in the number of olfactory receptors as we age. Although such receptors can be replaced—particularly in younger mammals—the rate of replacement falls with age, resulting in smaller olfaction area in the nasal cavity and a reduced number of receptors within that area. Much the same process—although not as severe and generally less noticeable—occurs to taste buds on the tongue. For example, as young adults we typically have only half as many taste receptors as when we were children, and the loss continues as we age. Even so, as we age, our ability to discriminate between the five major tastes—sweet, sour, bitter, salty, and umami—diminishes much less than our ability to distinguish odors. Because odor plays a major role in our ability to savor food, we have a severe subjective loss in the ability to enjoy what we eat or to discriminate between highly palatable and uninteresting foods.

The gradual loss of taste and olfactory receptors and of the ability to replace lost receptors can be attributed to cellular aging. While there are no current treatments for these related sensory losses, re-extension of the telomeres of the remaining receptors is promising as an effective intervention.

Hearing

Even in this age of texts and emails, our ability to hear remains critical to most social communication. Our culture assumes and all but requires the use of sound in order to function, sign language and handwriting notwithstanding. As we age, the ability to hear becomes impaired—a common, clichéd image of old age is the elderly person with one hand to an ear, asking, "What?" Presbycusis—age-related hearing loss—is almost universal, though its severity is widely variable. Curiously, the most common age-related hearing loss is not of the ability to hear very soft sounds, but of the ability to hear high-pitched sounds, which makes it difficult to distinguish consonants in others' speech—or to hear birds chirping or a phone ringing in the next room.

So as we age, it becomes harder to understand speech, whether the speaker is in the room with us or a character in a movie or TV show. This has little impact on low tones—as with vowels—but a dramatic impact on high tones—as with consonants. It's not that we can't hear that someone is speaking. It's that it becomes harder to tell the difference between, for example, *bed* and *bet*, *feed* and *feet*, or *rack* and *racked*, and thus harder to decipher the meaning of words strung together in sentences.

Our fading ability to catch the higher-pitched consonants is due to the way our auditory receptors work. Sounds consist of transient standing waves (see chart p. 120), which trigger signals from the auditory receptors, shown here as black circles. High-frequency sounds trigger more receptors; lower frequency triggers fewer receptors. As we age—and as we lose some of our auditory receptors, shown here as empty circles—we lose the ability to distinguish high-frequency sounds, including consonants in speech.

While the fading ability to hear high pitches (and understand speech) affects nearly all those who grow old, many of us encounter other, more general kinds of hearing loss as we age. Generalized hearing loss can have any number of causes—widespread loss of auditory receptors, nerve damage, atherosclerosis, diabetes, hypertension, trauma, drug damage, and others. While many of these other causes, such as atherosclerosis, are linked directly to cell aging, others, such as trauma or drug damage, are unrelated to aging. The loss of receptors, however, is directly caused by cell aging, as auditory receptors aren't replaced and don't function as well. Once again, the only potentially effective intervention would be to re-extend the telomeres of the remaining cells.

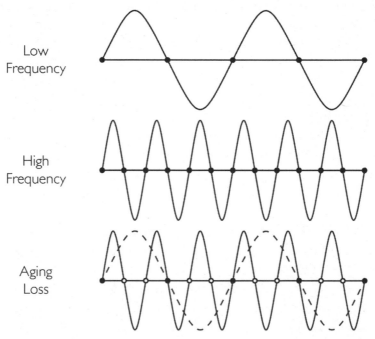

Hearing loss due to aging.

Vision

Most of us think of vision as the most indispensable sense. Vision changes as we age in many ways, although complete blindness is uncommon. After age forty, almost everyone finds it increasingly difficult to focus on nearby objects. We have difficulty reading and doing close work like threading a needle or tying on a fishhook. Although this change—presbyopia—is usually attributed to changes in the aging lens, it may also be related to changes in the ciliary muscles, which control the shape of the lens, as well as changes in the shape of the eyeball and associated structures (astigmatism). The lens itself is a collection of transparent cells that refract light, creating a focused image on the retina, and it can focus on near or far objects. These lens cells have no direct blood supply or any mitochondria, but remain metabolically active. The principal transparent protein, crystallin, is produced by these cells, but it is unclear to what degree these proteins are recycled in the adult lens.

Although there is some disagreement about the major cause of age-related presbyopia, it may be the gradual accretion of cells to the outer layers of the lens, making it less elastic and altering its shape. After age twenty, the lens becomes rounder, requiring the ciliary muscles to work harder to bring objects into focus. If this simple model is accurate, then cell aging per se may not play a significant role, and we are left with the time-honored approach to treating presbyopia: glasses or contact lenses. On the other hand, it remains entirely possible that epigenetic changes indirectly effect changes in the turnover of lens proteins or in the shape of the lens itself. And it's equally possible that aging in the cells of the ciliary muscles weakens their ability to focus the lens, meaning that these cells could be an effective point of intervention if we can re-extend telomeres. Whether or not telomerase therapy can mitigate presbyopia remains an open question.

A more subtle visual change with age is the gradual loss of contrast sensitivity, which happens as our retinas gradually lose the ability to detect fine detail. This may be due to a number of factors, but the loss of retinal ganglion cells is probably a major cause. Ganglion cells are responsible for the initial processing of

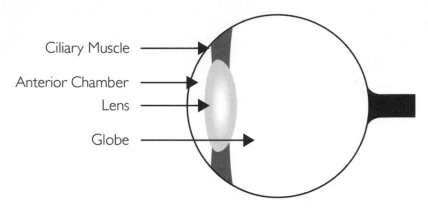

Ciliary Muscle

Anterior Chamber

Lens

Globe

Ciliary muscles control the shape of the lens in the human eye, enabling it to focus on objects near and far. Opinions vary as to why it becomes difficult to focus on close-up objects as we age.

visual information within the retina, before this information is sent to the brain. Ganglion cells are tuned to different frequencies of detail, so that as these cells diminish in number, we lose some ability to distinguish fine details. It's kind of like going from watching a high-definition TV to an old-fashioned standard-definition model. Curiously, a subset of these cells is also responsible for the pupil's response to light and for the modulation of our circadian rhythms, both of which present problems as we age, probably due to a gradual loss of these cells.

Presbyopia is the most common age-related visual change, but it's not the most feared. Gross loss of actual vision, as opposed to mere acuity, can be the result of a number of diseases, including macular degeneration, glaucoma, cataracts, diabetes-related ocular disease, and others. While some of these are clearly related to vascular changes—which are addressed in the next chapter—others appear to be purely ocular in cause.

Age-related macular degeneration (AMD) is one of the leading causes of blindness, particularly in the elderly. It is infamous in attacking our "center of visual attention," the fovea, so that people find themselves unable to see whatever they are particularly trying to focus on, even if their peripheral vision is relatively unimpaired.

This makes it impossible to read, see faces, or do anything else that involves visual detail. People with AMD can still see peripheral objects, allowing them to navigate and perform many daily tasks, but the disability is pronounced and progressive. The risk is high and increasing in older people. Perhaps one in ten have at least some macular degeneration in the decade after their retirement (sixty-five–seventy-five), but the incidence climbs to one in three in the following decade. Macular degeneration begins with deposition of a yellow pigment, drusen, in the macula. The source of this pigment—perhaps local cells, perhaps immune-system cells—is uncertain, but we can reasonably infer that it is produced and modulated by cells and that these functions become disordered as cells age. If this is the case—that these deposits are due to epigenetic changes in aging cells—then we would expect current interventions to be ineffective, as indeed they are, because they are aimed at the effects rather than the underlying causes. Whether or not telomere re-extension would prevent or cure age-related macular degeneration remains unknown, but it would be a more likely point of intervention than any therapy now available.

Cataracts are a disease of the lens that becomes more common with age, causing about half of all blindness globally. While "aging" is usually listed as a major cause, cataracts are also linked to diabetes, trauma, exposure to radiation (particularly ultraviolet), genetics, skin diseases, tobacco use, and certain drugs. As cataracts progress, the lens becomes less and less transparent, finally making it impossible to see at all. The word itself is a metaphor referring to the white (and not transparent) water in a waterfall. Cataracts appear to be associated with a rise in the water content of the lens proteins. The proteins themselves become denatured and begin to degrade with age. Although the standard assumption has been that lens proteins do not change over your lifetime—and hence are not maintained or turned over—recent work suggests that this is an oversimplification.[7] Although the mechanisms are unclear, lens proteins do show gradual replacement over time, even in the adult,

---

[7] http://www.ncbi.nlm.nih.gov/pubmed/23441119

and this is apparently the result of changes in gene expression. The lens—even the adult lens—is a dynamic organ that normally maintains the ability to transport and replace crystallin protein. The failure of this ability causes cataract formation. Whether normalization of epigenetic patterns and a return to normal protein maintenance can prevent or mitigate cataracts remains unknown, and hope depends upon clinical trials.

Glaucoma is sometimes called the "silent thief of sight," although the rapid-onset "closed angle" form is usually immediately painful and thus far from silent. It is the second-leading cause of blindness globally, after cataracts. Glaucoma is usually due to increased pressure in the anterior chamber of the eye—the space in front of the lens and behind the cornea. This pushes the lens back, increasing the pressure within the globe itself, which slowly compresses the blood supply to the eye. The result can be death of the retina cells, the visual receptors, and the optic nerve, ultimately causing blindness. The cause of the problem lies in the resorption of fluid—the aqueous humor—from the anterior chamber. As fluid continues to be produced without being resorbed, pressure rises and glaucoma ensues. Current medical therapy emphasizes either decreasing the production of aqueous humor or increasing the outflow through various mechanisms. While there are at least two major types of glaucoma, open-angle and closed-angle, the precise cellular mechanisms and their relationship to cell aging remain all but unexplored. In several cases, however, genes in the telomeric region have been linked to certain forms of glaucoma. In the more typical forms of glaucoma, the relationship to aging itself certainly suggests that epigenetic changes underlie the disease and respond to telomerase therapy.

Direct Telomere-Associated Diseases

Aside from normal age-related diseases, there are several diseases that have been shown to be directly related to telomere maintenance. Many of these are due to abnormal function of telomerase or the telomere itself and are generally genetic.

Dyskeratosis congenita

Dyskeratosis congenita (DKC) is a genetic disease caused by abnormal telomere maintenance, specifically due to one of at least three (and probably more) mutations affecting the RNA component of telomerase. As a result, cells—and especially stem cells—are unable to maintain normal telomere lengths during development. As might be expected, the chromosome with shortened telomeres has an increased risk of genetic mutations and thus the risk of cancer. In addition, DKC patients have unusual skin pigmentation, premature graying of the hair, abnormal nails, and other features. The problem is typically discovered prior to puberty, and about 75 percent of the patients are male. The major clinical problem is not the obvious signs, but the fact that the bone marrow is abnormal in 90 percent of these patients, causing death in about 70 percent, usually from bleeding or infection and sometimes from liver failure. Predictably, given the telomere abnormalities, many of the features of DKC somewhat resemble early aging. There is good reason to think that telomerase therapy would be effective in treating DKC, particularly because the problem can be reversed in the laboratory if we can reset telomere lengths.[8]

Progerias

Progerias include a number of related syndromes, including Hutchinson-Gilford progeria, Werner's syndrome, acrogeria, metageria, and others. In the case of "classic" or childhood-onset progeria (Hutchinson-Gilford), a rare genetic abnormality, telomeres are short at birth, apparently related to a defect in Lamin-A, a protein that affects the internal nuclear membrane and results in abnormal telomere maintenance. Telomere lengths at birth correspond to telomere lengths found in normal people in their seventies or

---

[8] Gourronc, F. A. et al. "Proliferative Defects in Dyskeratosis Congenita Skin Keratinocytes Are Corrected by Expression of the Telomerase Reverse Transcriptase, TERT, or by Activation of Endogenous Telomerase Through Expression of Papillomavirus E6/E7 or the Telomerase RNA Component, TERC." *Experimental Dermatology* 19 (2010): 279–88.

older. Given their "old cells," it is unsurprising that these children look old and die at an average age of 12.7 years, generally of atherosclerosis—heart attacks and strokes. While they appear old and have "old" blood vessels, skin, hair, and joints, they do not have a correspondingly high risk of Alzheimer's disease or immune senescence. The Lamin-A defect results in an abnormal protein—progerin—which is also seen in normal aged cells. Although this abnormality elicited considerable attention initially, attempts to correct the abnormal protein using farnesyltransferase inhibitors have been unrewarding, suggesting that the treatment may be simply locking the barn door after the horse has been stolen. The more obvious intervention—re-extension of telomeres—has not been pursued, despite being suggested twenty years ago[9] and since then in medical textbooks.[10]

HIV/AIDS

HIV, or AIDS, has a curious relationship to more common age-related diseases, because rapid cell turnover results in drastically shortened telomeres in the lymphocytes. The outcome is evocative of severe immune-system aging, albeit restricted to particular cell types and with the addition of whatever other damage the virus inflicts on the organism. The HIV virus infects cells of the immune system, especially T cells and dendritic cells. As the infected cells die and as the body responds by cell division to create more immune cells to replace them, the telomere lengths among these cells become shorter and shorter as the disease progresses. As long as the body can continue to replace the missing lymphocytes, there is an unstable balance, but as the telomeres shorten further, the rate of cell division slows and the functional ability of the replacement cells falls as well. There is then a relatively sudden inflection from morbidity to mortality. There has long been a proposal that telomerase, while not a cure for HIV, might well

---

[9] Fossel, M. *Reversing Human Aging* (New York: William Morrow and Co., 1996).
[10] Fossel, M. *Cells, Aging, and Human Disease* (New York: Oxford University Press, 2004).

prevent death by allowing the body to mount an indefinite immune response, rather than running out of immune cells. Because there are several fairly effective treatments for HIV—antiviral agents, highly active antiretroviral therapy (HAART), and particularly the HIV protease inhibitors—as well as the anticipated HIV vaccine, the possible benefits of using telomerase have been given short shrift, but remain a potentially effective point of intervention.

Cancer

Cancer is a major issue in any discussion of telomeres as a therapeutic target. Although the question arises in many different forms, it always centers on the safety of telomerase or the more complex question of whether long telomeres protect against cancer or promote it. Put succinctly, does telomerase cause cancer?

No, telomerase does not cause cancer.

Yes, telomerase may well prevent most cancers.

The discussion begins with a kind of paradox: Most cancer cells express telomerase, which should lengthen the telomeres, and yet most cancer cells have short telomeres. In normal cells, on the other hand, which don't usually express telomerase, those with the longest telomeres are the least likely to become cancerous in the first place. The presence of telomerase in cancer cells would suggest that telomerase might be detrimental, but the fact that long telomeres are protective suggests that telomerase might be beneficial.

Moreover, while a telomerase inhibitor is considered a good candidate as a cancer therapy, the use of a telomerase activator is likely to protect against cancer. In other words, telomerase *inhibition* could be used to treat cancer, but telomerase *activation* could be used to prevent cancer.

How can both of these statements be true?

To understand the value of telomerase in either cancer cells or normal cells, we need to understand some things about cancer itself. Most cancers occur due to an abnormal gene, or often several genes, whether genetically inherited, which is rare, or the more common result of an acquired mutation. There are many inherited genes that don't necessarily *cause* cancer but are certainly known to increase its likelihood. Perhaps the most widely known of these is found in women with an abnormal *BRCA1* or *BRCA2* gene, many of whom (depending on the precise gene) have an 80 percent chance of breast cancer. This particular set of genes is involved in DNA repair, so an abnormal BRCA gene increases the risk of DNA damage, because acquired damage can no longer be repaired. In cancers generally, we see a similar pattern of increased problems in maintaining normal genes and normal chromosomes—genomic instability. In cancer, whatever the source of the abnormality—inherited, acquired, or both—the cellular result is usually genomic instability, and the clinical result is that sooner or later the cell becomes cancerous.

Metaphorically, we might think of cancer cells as sociopaths lurking in the cellular neighborhoods of their tissues. Normal cells have specific local functions and are restricted from abnormal division by both internal and external factors. In normal tissues, cells receive chemical signals telling them when to divide and when not to. For example, normal breast cells don't divide unless replacement cells are needed, but cancerous breast cells divide regardless of such signals. Normal, noncancerous cells that have DNA damage have internal signals that keep them from dividing even when external signals tell them to do so. Cancer cells, on the other hand, ignore all inhibitory signals—whether internal or external, whether there is DNA damage or not—and begin to divide regardless.

The problem with cancer cells is twofold: They multiply, and they don't work. Frequent cell division results in a mass of unnecessary cells, and these cells not only use up metabolic resources, but the mass itself can be fatal, as may occur in brain cancers, where the growing mass can squeeze the brain within the confines of the skull. Moreover, the fact that cancer cells don't respond to internal or external signals—which is why they divide when they shouldn't—affects many other cell functions. Cancer cells that are supposed to create a specific protein—as are some white cells—might create the wrong protein, too much of the right one, or none at all. Of course, if a single damaged cell behaved that way, it would cause no real problem as one bad cell among a million normal ones. The reason that cancer cells are so deadly is that they keep dividing until there are a great many of them. In short, the key to preventing cancer—from a clinical perspective—is to prevent inappropriate cell division.

Every normal cell in your body has three major protections against inappropriate cell division due to DNA damage. The first is that the cell detects and repairs damaged DNA. Cancer cells are incapable of repairing their own DNA damage, so the first line of protection fails. The second line of protection is that whenever DNA damage is detected, a normal cell shuts down the cell cycle so that it can no longer duplicate the damaged DNA and divide. Even if the DNA never gets repaired, at least the cell can no longer multiply and create more damaged (i.e., cancerous) cells. This cell-cycle braking system is enormously effective. Consider what happens, however, if part of the DNA damage is actually damage to the cell's braking system itself. In this case, the damaged cell continues to divide regardless of the risk to the organism. The cancer cells begin to accumulate and grow more numerous.

Telomere shortening—and cell aging—is the third line of protection against cancer.

Telomeres work in two ways. As telomeres shorten, they begin to deactivate the cell via epigenetic changes, and even if that fails, the loss of the telomeres ultimately guarantees that the chromosomes will become inoperative as they begin to lose not only telomeres, but (in the extreme case) genes themselves. Cells with shortening telomeres are increasingly likely to die because of increasingly severe epigenetic changes, but cells with no telomeres cannot survive because of the wholesale loss of genes. In order to survive, cancer cells must maintain telomeres, at least minimally, and so they do.

Most cancers manage to maintain short but barely adequate telomeres, with the result that they manage to survive and multiply, but their shortened telomeres cause a very high rate of mutation. Cancer cells are continually mutating, and although many of them die, those that survive are increasingly resistant to inhibitory signals as well as most other limits placed on normal cell growth.

In essence, once the cell escapes growth limitations and begins to mutate, its daughter cells become more and more malignant, simply because they are selected for those characteristics. This same process continues as cancer cells may come to evade the body's immune defenses: The rapid rate of mutation ensures that cells that survive do so because they are capable of evading such defenses.

One might reasonably wonder how a cancer can possibly survive through all the internal and external defenses and how it can survive if it is mutating, because mutation implies damage to the cancer cell. The answer is that most cancer cells do not survive. Most early cancer cells stop dividing in response to either internal or external inhibitory signals. Those that survive that stage have shortening telomeres, and almost all of those cells die of cellular aging or wholesale chromosomal damage. Those that survive by managing to maintain some minimal telomere length often die in response to the body's immune response or—in the case of solid cancers—because they can't maintain an adequate blood supply. Most cancer cells don't survive, which is why most of us don't die of cancer at an early age.

The problem with cancer is that not all cancer cells succumb to our defenses. The very few that manage to evade all these obstacles are quite enough to cause the cancers that kill so many of us.

Then what is the role of telomerase in cancer?

If the telomeres are long enough, cells have genomic stability; they can prevent and repair genetic damage efficiently and have a high probability of not becoming cancerous. So if a cell can express enough telomerase to keep long telomeres, telomerase is protective against cancer.

But if the telomeres are short, cells have genomic *instability*; they are unable to prevent further genetic damage and mutations and have a high probability of becoming cancerous. The problem is that many cancer cells produce just enough telomerase to maintain very short telomeres (or find an alternate way of maintaining such short telomeres). Putting it simply, it would be best for patient survival if our normal cells produced a lot of telomerase to protect against cancer *or* if cancer cells produced no telomerase at all and died quickly. The worst possible scenario is to have just enough telomerase to enable cancer cells to not only survive but to get worse over time, which is precisely what most cancer cells do.

This is why one of the promising avenues to treat cancer is to use a telomerase inhibitor, to cause cancer cells to die of cell aging. This would have the unfortunate side effect of also inhibiting telomerase in our stem cells, which are important to our long-term survival, but that may be a small price to pay in order to cure the cancer. The benefit is acute survival; the risk is chronic tissue failure and a high risk of various age-related diseases.

Given what we know of cancer, would telomerase be beneficial to normal people or not? In most cases, telomerase should be enormously beneficial in preventing or curing most age-related diseases as well as significantly lowering the risk of cancer. If the patient already has a cancer, however, the result is less clear. Extending telomeres in a patient with cancer might well be beneficial in that the cells may once again be capable of repairing their DNA damage and reversing early cancerous changes. In this scenario, telomerase might not only prevent but cure many cancers. On the other hand, if telomerase therapy only maintained shortened telomeres rather than re-extending them, then we would merely be maintaining the cancer cells. Another consideration is that restoring telomere length in a cell that has an *inherited* defect in its ability to repair DNA will only enable that cell to divide and pass on the defect; all the telomere length in the world won't fix the problem.

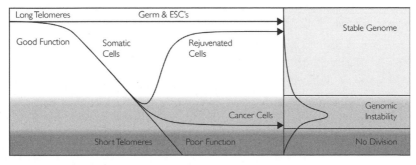

Telomere length and cancer.

What's really interesting is that while the human incidence of cancer climbs exponentially with age, the curve is exactly the same for mice and rats, even though they have lifespans perhaps only one thirtieth of ours. What this implies is that cancer mutations are *not* simply a matter of cumulative exposure to cosmic rays, UV radiation, and spontaneous molecular changes (molecules can isomerize at body temperature), because the rate of exposure is the same for mice and humans. Rather, they are a matter of a cumulative *decrease* in DNA repair, which is controlled by telomere shortening. In short, if we can reset telomeres, we can reset the incidence of cancer. We might use telomerase inhibitors to *treat* some cancers, but we can probably use telomerase activation to *prevent*—and I mean all but eliminate—most cancers in the first place. Telomerase—if sufficiently active—is very likely to prevent or cure some early cancers (rather than making cancer worse), but it won't help those with an inherited genetic problem.

The bottom line is clear: Telomerase is generally protective against cancer.

Summary

In the age-related diseases we have discussed, the clinical problems of aging result from cell aging. In cells that divide, telomeres shorten, the pattern of gene expression changes, and the cells become less and less efficient, as well as slower at replacing other lost cells. When old cells don't work and aren't replaced efficiently, aging tissues become dysfunctional, and we see the onset of obvious age-related disease.

If we can reset telomere lengths in such cells, we can reset the pattern of gene expression and make these cells functionally young again. As we will see, we have every reason to think that telomerase can be used to prevent and cure most age-related diseases.

CHAPTER SIX

⁓

Indirect Aging:
Innocent Bystanders

S o far, I've focused on aging cells and the diseases that arise in the tissues they make up. I call this "direct aging," a process of cause and effect that begins as telomeres shorten with each cell division and progresses through changes in gene expression and cell dysfunction to clinically obvious age-related disease in the similar, surrounding cells.

However, there are types of cells in our bodies that either never divide or that divide so infrequently that their telomeres shorten very little in the adult, whatever their age. One might think that this would protect non-dividing cells from age-related disease.

It doesn't.

The reality is entirely different. Some of our most prevalent and fatal age-related diseases affect cells that *don't* divide, because non-dividing cells are always critically dependent on cells that *do* divide. The majority of people on this planet now die because of diseases in which non-dividing cells—cells that don't age—fail because they are dependent on dividing cells that do age.

In the case of heart attacks, for example, we die because our heart muscle cells die. Those muscle cells themselves show little if any significant age-related change, but the survival of the cardiac muscle cells is completely dependent on their blood supply, which is delivered via the coronary arteries. Heart attacks occur because the coronary arteries are blocked. Cells lining the coronary arteries show rapidly shortening telomeres, paralleling and preceding the progress of atherosclerosis. We don't die of an old heart, we die of old arteries. One way or another, this kind of indirect pathology is the case with most age-related fatal diseases.

The diseases we fear the most are those that result from "indirect aging." "Innocent bystander" cells, which have no inherent dysfunction as we get older, die as a result of their dependence on other cells that are aging rapidly. In heart attacks, strokes, Alzheimer's disease, Parkinson's disease, and a host of similar diseases, the cells that die are dependent on other cells that have shortened telomeres.

Let's explore the two major categories of indirect aging: age-related arterial disease and age-related neural disease.

Cardiovascular Disease

The horror of cardiovascular disease—particularly myocardial infarction—is the suddenness of its attack. One moment we feel safe, healthy, and only marginally willing to concede our gradual aging; the next moment there is pain and terror, perhaps even sudden death. The vascular system takes decades to age, but we are often totally unaware of our growing risk, while the clinical outcome—the instantaneous descent into helpless mortality—hits us with an unexpected but inescapable reality.

While the term "cardiovascular disease" is common, we are really discussing a *primary* problem within the blood vessels (almost always the arteries), which then causes a *secondary* problem in the heart, the brain, or other end organs. First the vessels become diseased, and then the end organs fail. We might more accurately just call it vascular disease, but because it is the failure of the end organ that puts us the hospital, we expand the term to include it—in this case "cardiovascular aging." Of course, this term leaves out the brain and many other parts of the body that depend on healthy arteries. We may die because of old vessels, but it is the heart, the brain, the kidneys, or even our extremities where we see tragic problems as arterial aging progresses and ultimately proves fatal.

The general term for aging or "hardening" of the arteries is arteriosclerosis, but because cholesterol plaques are often part of vascular aging, the term atherosclerosis will serve here. Whether plaques form or not, arterial walls show changes with age, usually becoming less elastic and pliable as the cells fail to maintain the extracellular proteins—particularly elastin and collagen—that are necessary for normal, healthy vascular function.

As a result, our arterial walls become "hard" and lose their ability to stretch and respond to changes in blood pressure. The upshot is that aging arteries are more likely to form aneurysms that might tear or leak, the overall blood pressure gets higher, and blood pressure is less adaptable to changes in position or to changing physiological needs. Even though the measured arterial pressure often rises, the blood flow to end organs often falls, leaving the brain, for example, with higher pressure but a less adequate arterial supply. The higher pressure and loss of elasticity cause a growing risk of arterial wall ruptures and therefore of hemorrhagic stroke. Some ruptures are large, resulting in obvious weakness, an inability to speak, or rapid death; but many are tiny and cumulative, resulting in a gradual loss of brain function over decades, often called multi-infarct dementia. This same problem can occur in other organs as well, resulting in cumulative damage throughout the aging body.

When plaque formation occurs as a result of arterial cell aging, several other risks become evident. Over time, gradually increasing occlusion of the artery will cause ischemia and failure of the end organ (often the heart), unless the vascular system provides another route to supply blood to it (called "neovascularization"). Worse yet, the plaque may break loose, move down the artery, and instantly and totally occlude any arteries beyond where it lodges. When the clot blocks the blood supply to vital heart muscles, the result is a sudden heart attack, often causing immediate death. If the clot is in the carotid arteries and goes to the brain, large areas of the brain suddenly lose their blood supply. The result is an occlusive stroke with loss of brain function, often including paralysis of one side of the body or aphasia (inability to speak). Elsewhere in the body, a clot may cause tissue death in almost any vital organ, a kidney, the intestines, etc.

Most people—and certainly most physicians—feel that they understand the causes of atherosclerosis. After all, there is clearly a strong link with the Big Four risk factors—smoking, high blood pressure, high cholesterol, and diabetes—isn't there? In reality, there are some interesting and enlightening exceptions to this correlation. Some people may have some or all of these risk factors, yet have no atherosclerosis at all, much less suffer heart attacks. On the other hand, some people have none of the four risk factors, yet die of overwhelming atherosclerosis or the diseases we associate with it. In fact, as many as half of all heart-attack patients may lack any of these four classic risk factors. The best example is children with progeria. Almost none of these children have any of the four risk factors, and yet almost every one of them has severe atherosclerosis, and they almost all die of either a heart attack or stroke. How can they have the disease without any of the risk factors?

What is going on?

Does this mean our understanding of the disease is wrong? No, it simply means that our understanding is incomplete. For example, the simple model that high cholesterol directly causes cholesterol deposits in our arteries is simply inaccurate. Obviously, there must be other pathways that cause atherosclerosis. The disease may correlate with the classic risk factors, but something much more complex is going on. If that's the case, and the data shows that it is, then how does atherosclerosis *actually* work, and why do we still regard smoking, hypertension, high cholesterol, and diabetes as legitimate risk factors?

To understand the relationship between risk factors and disease—such as the relationship between serum cholesterol and atherosclerosis—we need to understand how the cells of the arterial walls age. I return to the metaphor in Chapter Two and Chapter

Five of the lake with hidden rocks beneath the surface. The effects of risk factors may lie well beneath our keel when we are young. In the case of smoking, for example, it may take decades for the damage to accumulate, but it is also true that young cells are better able to repair the damage smoking causes. As we age, as our telomeres shorten, and our cells become less functional and less capable of self-repair—the lake level drops and we begin to strike the hidden rocks we once sailed over safely. At age twenty, our cells can repair most of the damage caused by our risk factors, but by middle age and beyond, we can no longer keep up with the arterial damage caused by our tobacco use, hypertension, high cholesterol, or diabetes. In short, as our cells age, damage begins to accumulate. As cells no longer function properly or are not replaced fast enough, the arterial wall becomes stiffer and much more prone to rupture, and cholesterol begins to accumulate in plaques—and the vessel fails.

Even when we include the entire panoply of known risk factors—diet, alcohol, obesity, lack of exercise, high homocysteine

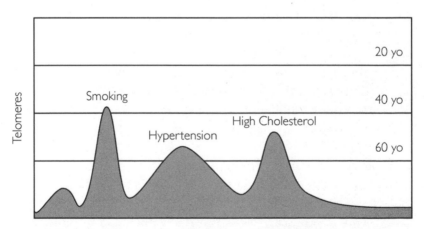

When we are young, our arterial cells adequately repair damage done by common risk factors. However, as our cells age due to shortening telomeres, they can't keep up with damage repair and replacement of dead cells. Like rocks in a lake where the water level is falling, cell damage gets closer to the surface, and we are more likely to "run aground" on a heart attack or stroke.

levels, individual cholesterol fractions and ratios, apolipoprotein E4, estrogen levels, tocopherol levels, prothrombotic mutations, elevated monocyte counts, C-reactive protein, myeloperoxidase, stress, dental infections, other bacterial or viral infections (such as herpes, cytomegalovirus, or coxsackie), and inflammatory biomarkers in general—we still find that risk factors alone do not adequately account for arterial aging. All of these risk factors are valid, but they can often be "hidden" by long telomeres in young cells.

The hidden rocks analogy is particularly apt in regard to progeric children, who start life with short telomeres. The cells that line their arteries are already old, unable to deal with even the most minimal of risk factors. While these children don't smoke and don't have hypertension, diabetes, or high cholesterol, what they *do* have is old cells—cells that cannot maintain extracellular elastin or collagen and cannot replace lost cells. And so their blood vessels accumulate cholesterol, even when the serum levels are normal, and rapidly show atherosclerotic changes.

These children die of heart attacks and strokes often even before age ten and with none of the characteristic cardiovascular risk factors. Short telomeres are more than enough.

The opposite may well occur in people who have some of the usual four risk factors but no obvious sign of atherosclerosis. They likely either have longer telomeres or are lucky enough to have genes that mitigate the normal risks. Returning to our metaphor, this is like saying that for such people the rocks are much smaller, so that the water level has to fall much further before the rocks become hazards.

Progeric children on one hand and those who seem impervious to risk factors on the other lead us to this conclusion: The Big Four risk factors—and all the other ones we know of—are real components of atherosclerosis, but don't give us anything like a complete understanding of the disease. To prevent atherosclerosis, we need to understand not merely a few risk factors, but cell aging in the arterial walls.

The arterial wall has several layers, although in the smaller, peripheral arteries, these layers become much simpler and thinner until only a single layer of cells is found in the walls of capillaries. The innermost cells—endothelial cells—receive the most damage over time, by direct exposure to toxins and other materials and from "shear stress." Just as river banks erode the most where the current is most forceful, the shear stress on the arterial walls is greatest along curves and wherever the arteries divide. As a result, these are the cells that are most frequently lost and most frequently replaced. The outcome is predictable: The endothelial cells lose telomere lengths and show a decreasing ability to function normally as we age. Rapid cell aging from toxins and shear stress is most marked in people with hypertension or diabetes and in those who smoke. In every case, there is a strict correlation between the loss of telomere length and the onset of arterial disease: Wherever we see atherosclerosis, we also see short telomeres in the endothelial cells.

As the endothelial cells begin to fail, they are less able to maintain the layers below them, particularly the elastic and other fibers. Also, they begin to separate slightly, giving toxins, viruses, and bacteria better access to subendothelial layers. The first grossly visible change then occurs as circulating monocytes and platelets begin to adhere to the endothelial wall, even before the lower layers show changes. But as endothelial failure progresses, the result is increasing inflammation in the subendothelial layers, as macrophages and other immune system cells begin to enter the arterial walls. Scarring follows soon after, and cholesterol begins to accumulate on the scar tissue. This buildup causes the wall to bulge, eventually blocking the artery and also increasing the risk that the lesion will tear loose and form a clot downstream.

While the changing dynamics of the cells—beginning with the aging of the endothelial cells—explain most of what happens in atherosclerosis, we cannot underestimate the effects on the fibrous layers that lie between the endothelial cells and the smooth muscle cells, as well as the outer layer of adventitial fibers. Aging cells fail to maintain the elastic and collagen fibers that provide the flexibility and toughness of the arterial walls. In young adults, the elastic nature of the large vessels, such as the aorta, allows them to stretch during the contraction of the heart (systole), and rebound as the

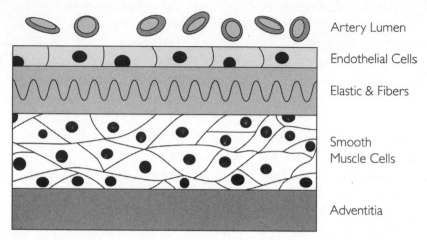

Artery Lumen

Endothelial Cells

Elastic & Fibers

Smooth
Muscle Cells

Adventitia

Structure of the artery wall: Endothelial cells make up the innermost layer next to the artery lumen—the open canal through which the blood flows. Adventitial fibers make up the outer layer.

heart refills (diastole). This elasticity evens out the sharp waves of pressure, relieving most of the chaotic shear force that damages the endothelial cells. As young arteries age, the failing endothelial cells no longer maintain the elastic fibers, so that blood pressure becomes more damaging, and endothelial cells fail all the more rapidly.

The extent of changes in the aging endothelial cells is disconcerting. They lose mitochondria and show general deterioration. The endothelial tissue becomes thin, irregular, and occasionally missing altogether. It fails to act as a barrier, fails to regulate arterial blood pressure, and becomes less responsive to vasodilators (which your body uses to control blood pressure). Not only does blood pressure become more of a problem, but the more peripheral organs begin to lose blood flow.

Curiously, this cascade of pathology does not begin in the intermediate layers, despite the fact that the changes there are more obvious. The pathology in the subendothelial layers—fatty streaks, calcification, cholesterol deposition, inflammation, smooth muscle proliferation, foam cells—is all secondary, both in time and in causation, to the aging changes in the endothelial cells. This cascade from endothelial cells to the subendothelial layers also explains why none of the risk factors is absolute. Any process that can increase the aging of the endothelial cells will trigger the overall disease, but unless the endothelial cells lose telomere length, none of the risk factors will necessarily trigger the disease. Our conventional understanding of atherosclerosis has been incomplete. It isn't a case of "high cholesterol causes heart attacks" (or of diabetes, smoking, or hypertension causing heart attacks). The actual process is more subtle and complex.

On the other hand, the very nature of this subtle and complex cascade makes clinical intervention clear and simple: *re-lengthen the telomeres*. No therapeutic intervention, no matter how well-intentioned, can hope to cure or prevent arterial disease merely by addressing a more-or-less distant risk factor like hypertension or smoking. But if we address telomere length in the endothelial cells, we can bypass—and perhaps all but ignore—most of the common risk factors we currently associate with arterial disease. The importance of cell aging—over any other risk factors—is seen in arterial stents in which antisense nucleotides are used to block cell division, with the result that restenosis (which frequently happens otherwise) no longer occurs, even in individuals with high-fat diets.

In the laboratory, resetting telomere length has been shown to reverse the age-related changes in human endothelial cells and tissues.[1] While numerous studies support the physiological efficacy of re-lengthening telomeres, clinical studies have not yet advanced to the point of testing similar interventions in actual patients. As we will see, however, even clinical studies point to the probable efficacy and we now have the technical ability to do human trials using agents to reset telomeres in endothelial cells.

The most promising single approach to the prevention or cure of age-related arterial disease—including atherosclerosis and myocardial infarctions—is to re-extend telomeres within the endothelial cells of human arteries.

---

[1] Matsushita, S. et al. "eNOS Activity Is Reduced in Senescent Human Endothelial Cells." *Circulation Research* 89 (2001): 793–98.

CARDIOVASCULAR DISEASE: QUICK FACTS

Age: Arterial changes occur even in young people, particularly those with multiple risk factors (e.g., tobacco use and dietary risks), with many young adults (especially in developed countries) showing prominent arterial disease in their twenties and thirties. The changes accelerate in men during their fifties. In women, the disease may lag behind as much as ten years, although it rapidly catches up after menopause. The most common age for the first heart attack, especially a fatal heart attack, is fifty-five to sixty-five, but the risk then remains high and the risk of fatality climbs steadily thereafter.

Cost: Coronary atherosclerosis alone is among the most expensive causes of inpatient hospital admission in the US. Annual costs exceed $10 billion.[1]

Diagnosis: Diagnosis of heart attacks may be clinical, but when the event is nonfatal, it is confirmed via EKG changes, enzyme levels in the blood, or a radiological procedure such as a coronary artery study. Diagnosis of arterial disease per se is generally made through radiological studies such as arteriography.

Treatment: Prevention generally stresses exercise, diet, and smoking cessation. Treatment also includes pharmaceutical and surgical interventions. Currently, the most commonly prescribed pharmaceuticals are statins, although niacin, so called cholesterol-lowering agents, and other drugs still play a role. Surgical approaches include coronary artery bypass and coronary artery stents.

---

[1] https://www.hcup-us.ahrq.gov/reports/statbriefs/sb168-Hospital-Costs-United -States-2011.jsp

Carotid Artery Disease

Heart attacks may terrify us, but strokes reach even deeper. Not only are we suddenly reminded of our own mortality, but we find our own brain has become a traitor. It's easier to accept the limitations imposed by a heart attack than those that result from a stroke.

We lose some or all of the use of our legs, our hands, or our speech. No longer can we walk, run, dance, write, play an instrument, cook, or tell others what we're thinking. Pieces of our humanity are—without warning or mercy—simply ripped away. The possibility worries us before it happens, the reality terrifies us once it does.

In most respects, carotid artery disease is simply a subset of arterial disease as we've already described it. The major differences are location and outcome—the complications are brain-related. Preeminent among these complications is the cerebrovascular accident (CVA), usually referred to simply as a stroke.

Strokes occur whenever the blood supply to the brain is interrupted, either through obstruction by a clot (thrombosis) or because of bleeding from the arterial supply (hemorrhage). Thrombotic strokes can be treated by resolving the obstruction, for example by dissolving the clot by the use of thrombolytics. Hemorrhagic clots, however, are very rarely amenable to therapy, whether medical or surgical.

Regardless of either the specific cause or the long-term prognosis, the immediate problem is exactly the same: As part of the brain loses its blood supply, it can no longer function. The acute symptoms may be an inability to move, speak, understand speech, or see, and sometimes the stroke is fatal. Because the two sides of the brain have a largely independent blood supply, in most cases the symptoms may be one-sided, such as the inability to move one limb or one side of the body. Patients with no history of trauma and who present with a one-sided paralysis of the leg and/or arm, for example, are considered to have had an acute stroke until it's proven otherwise.

The arterial pathology underlying a stroke is precisely the same that underlies a heart attack. In both cases, the arterial walls show characteristic changes and in both cases these changes are due to age-related changes within the endothelial cells, changes which result in increasing damage due to known risk factors and the inability of arterial wall cells to keep up with these forms of damage.

STROKES: QUICK FACTS

Age: Although strokes can occur at any age, about 75 percent of first strokes occur after age sixty-five, and the incidence doubles each decade.[1] The most important single risk factor, other than age itself, is hypertension, but this is closely followed by history of previous strokes, diabetes, high cholesterol, smoking, atrial fibrillation, hypercoagulation, etc.

Statistics:[2] Stroke is the third leading cause of death in the US and the second globally. It is the leading cause of significant long-term disability. Costs in the US alone are estimated at more than $40 billion annually.

Diagnosis: Initial diagnosis is almost always clinical, although symptoms can occasionally be misattributed to other causes. Diagnosis almost always includes a CT or MRI scan to evaluate whether there is bleeding (hemorrhagic stroke) as well as to identify the area of involvement.

Treatment: Prevention usually addresses blood pressure, smoking cessation, and control of atrial fibrillation (or anticoagulation). A thrombotic stroke can be immediately treated with thrombolytics or, more rarely, neurosurgery, but there are no generally effective therapies for strokes per se. Once the neurons have died, the damage cannot be undone; controlling risk factors (against future strokes) and stroke rehabilitation are currently the standards of care. Risk can often be lowered by control of hypertension, as well as through the use of antiplatelet medications, statins, anticoagulants, or occasionally with carotid endarterectomy.

---

[1] http://www.strokecenter.org/patients/about-stroke/stroke-statistics/
[2] http://www.strokecenter.org/patients/about-stroke/stroke-statistics/

Hypertension

Blood pressure tends to climb with age. This rise is partly a result of changes within the arterial walls themselves, as described earlier. But it also results from aging elsewhere in the body: the kidneys (which play a prominent role in setting blood pressure), the endocrine system, the heart, the brain, et al. Blood pressure is commonly measured in two major components: systolic and diastolic pressure. Systolic pressure, the pressure at the "top" of the cycle when the heart is contracting, is more volatile in its responses to stress, worry, body position, and other factors. Diastolic pressure, the pressure at the "bottom" of the cycle when the heart is refilling, is a bit more constant and less variable with transient factors. While some forms of hypertension are not strictly age-related, the majority of clinical hypertension is closely linked with age-related changes.

Hypertension not only causes additional work for the heart, but it increases arterial damage, kidney damage, and the risk of arterial aneurysm or rupture, including hemorrhagic stroke. While the underlying causes of age-related hypertension remain uncertain, there is a growing body of data suggesting that the key features[2]—increased peripheral resistance due to narrowing small arteries and a decreasing capillary bed—are the result of endothelial cell dysfunction, just as it is in the case of arterial disease in general. The aging of arterial endothelial cells in large arteries causes damage to accumulate in the remainder of the arterial wall (atherosclerosis). It causes smaller arteries to become narrower and less pliant. And it causes the loss of the smallest vessels, capillaries, altogether.

Perhaps counterintuitively, increasing hypertension does *not* increase blood flow to the end organs. In fact, hypertension in large vessels causes hypotension—low blood pressure—in the end organs. In hypertension, the smallest vessels either narrow (in the case of small arterial vessels) or disappear (in the case of capillaries). The result is that hypertension is measured in the doctor's office along with decreasing function in the end organs that are no longer getting enough blood flow, regardless of the blood pressure as it leaves the heart.

To make matters worse, in some cases the body's response exacerbates the problem. The kidneys, for example, attempt to regulate systemic blood pressure. That's part of their job. As hypertension progresses, the kidney cells are actually seeing less perfusion at the cellular level, so in order to increase perfusion, the kidneys respond

---

[2] Fossel, M. *Cells, Aging, and Human Disease.* Oxford University Press, 2004 (see Chapter Nine).

by increasing systemic blood pressure. In the long run, unfortunately, this merely increases the rate of aging of those endothelial cells lining the arterial vessels and capillaries. Small arteries get narrower and more capillaries are lost, and this increases blood pressure still further in a vicious cycle that ends variously in kidney failure, heart attacks, strokes, aneurysms, and other clinical tragedies.

Congestive Heart Failure

Congestive heart failure (CHF) is a collection of diseases with varying causes. In most cases, however, aging itself underlies the common outcome. Congestive heart failure is ascribed—reasonably enough—to a failing heart. The disease itself is often broken down into left-sided or right-sided failure, although this division is somewhat simplistic. The left side of the heart receives blood back from the lungs and pumps it to the rest of the body; the right side receives blood flow from the body and pumps to the lungs. In either case, one of the primary triggers for heart failure is myocardial infarction, with the result that the damaged portion of the heart muscle is no longer able to pump effectively.

Perhaps three-quarters of all congestive heart failure—including those cases resulting from myocardial infarction and hypertension—can be ascribed to cell aging, but a number of causes—including smoking, viral infection, valve disease, and others—are either not age-related or only distantly related to problems caused by cell aging.

Neurologic Diseases of Aging

A number of neurologic diseases are age-related. The most prominent one is Alzheimer's dementia (or Alzheimer's disease), but Parkinson's disease is almost as well known and widely feared. There is a hodgepodge of other age-related diseases and conditions, including loss of motor coordination, poor reflex function, age-related sleep disturbances, etc. Although such diseases and conditions have long been defined separately and considered to have distinct pathologies, there is a growing perception that they are a spectrum of diseases with a shared causation. Parkinson's disease manifests predominantly in the substantia nigra, while Alzheimer's attacks many locations but particularly the cerebral cortex and some subcortical structures. Yet the cause of cell death might be much the same for all of these neurological diseases, regardless of the part of the brain they afflict.

Nonetheless, we'll discuss Alzheimer's disease and Parkinson's disease separately.

Alzheimer's Disease

O the mind, mind has mountains; cliffs of fall
Frightful, sheer, no-man-fathomed. Hold them cheap
May who ne'er hung there. Nor does long our small
Durance deal with that steep or deep . . .

— Gerard Manley Hopkins

Of all the diseases of aging, Alzheimer's is the most terrifying.

Alzheimer's is the thief in the night, stealing our souls, leaving only empty shells behind as it staggers away. Many aging diseases may kill us, others limit what we can do, but Alzheimer's limits what we can understand. It robs us of our innermost selves, our minds, our intellect, our personal souls. It takes away our ability to be who we are. Every world literature has stories in which some dark force—the devil, a curse, black magic, dementors—can remove the soul and leave only a golem, a zombie, a husk. This is the horrifying reality of Alzheimer's.

Many people, knowing something about telomeres and nothing about human pathology, have suggested that cell aging cannot cause Alzheimer's dementia. They naively argue that because neurons (generally) don't divide, their telomeres can't shorten with age, so that cell aging can't possibly underlie Alzheimer's disease. As in the case of heart disease, however, the point is irrelevant.

Neurons may not age directly, but they suffer from the pronounced aging of the cells they depend on for survival. Specifically, the microglial cells in the brain do indeed age, resulting in defective

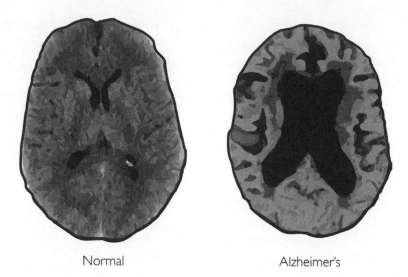

Normal   Alzheimer's

Cross-sections of a normal brain and one in an advanced stage of
Alzheimer's disease.

support for the "innocent bystander" neurons. The outcome is
Alzheimer's disease.

As with atherosclerosis, the pathology of Alzheimer's disease
is more complex than some scientists and physicians recognize. In
the case of Alzheimer's dementia, we need to understand the role of
beta-amyloid and tau proteins. It is true that these two proteins play
the role of the "evil minions" in the disease, but it is the microglia
that lie behind the entire tragedy and direct the attack. Microglia
show cell aging and it is this critically important actor that results
in the death of our neurons.

Glia make up about 90 percent of all the cells in the brain, and
microglia make up about 10 percent of all the glial cells and are
more commonly found near neurons. Microglia are "immigrants"
to the nervous system. Essentially immune cells, they invade the
brain from the bloodstream and take up residence around the neurons. When activated by injury or infection, they transform into
macrophages, divide, and attempt to deal with the problems. In
doing so, they divide repeatedly, shortening their telomeres, and

becoming dysfunctional. This is the first step in the progression of Alzheimer's.

These aging microglia are now less and less capable of maintaining the neurons, especially with regard to beta-amyloid production and turnover. Microglial cells become "activated" and change both their shape and their function, becoming more and more inflammatory, which accelerates damage. Acting together, the microglia and neurons begin to produce short, damaged beta-amyloid molecules—oligomers—which are toxic to the neurons. We begin to see larger deposits of beta-amyloid plaques as the damage progresses. As the neurons find themselves overcome by the growing damage, their tau proteins, which are critical to maintaining their axons and hence to carrying signals from one neuron to another, begin to accumulate in the neuron body. Finally, the inflammation, microglial failure, beta-amyloid toxicity, and tau protein tangles overcome the ability of the neurons to tolerate the damage, and they begin to die.

Alzheimer's disease gathers speed. First we forget our keys, then the names of our loved ones, and finally how to care for ourselves at all.

While there is a growing recognition that arterial aging also plays a role in—or is at least correlated with—Alzheimer's disease, most researchers have focused exclusively on nerve cells. In this narrow focus, they have ignored changes not only in the arteries but also in other structures, such as the blood-brain barrier, and other cell types, such as the glia. Historically, this narrow focus is understandable. The most obvious histological changes are seen in the cortical neurons, and these are the cells we most associate with our cognitive abilities. Moreover, we have long known that neuronal death is preceded by an accumulation of beta-amyloid protein around the neurons as well as an accumulation of tau proteins within those same neurons. Unfortunately, this all-too-obvious observation has resulted in numerous, costly, and uniformly failed attempts to cure or prevent Alzheimer's by aiming only at these two targets—beta-amyloid and tau proteins.

The results of clinical trials have been depressingly predict-able: Nothing has worked. There are more than 1,600 clinical trials with almost 500 still in progress. While some of these are meant to provide symptomatic relief (as with acetylcholinesterase inhibitors), many have the intent of changing the course of the pathology itself, slowing or even stopping the disease's progression. Many of these trials aim at the same two targets: beta-amyloid and tau proteins. That is understandable: both beta-amyloid and tau proteins are conspicuous parts of the microscopic pathology and are reasonably considered central to it. Beta-amyloid, for example, is important to neural function, but is known to be toxic when present in large amounts—such as the amounts that surround dying neurons in the brains of Alzheimer's patients. Tau proteins are likewise essential to the internal structure of neurons, but tangles of tau proteins fill those same dying neurons in the brains of Alzheimer's patients. Ergo, both these proteins are reasonable candidates for therapeutic

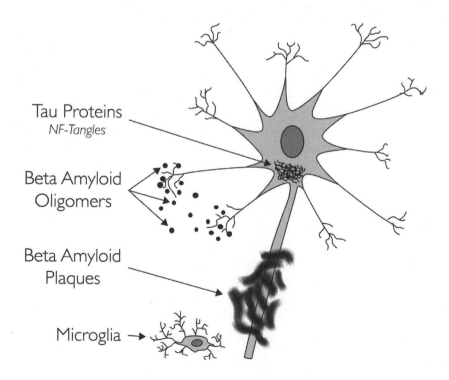

Tau Proteins
NF-Tangles

Beta Amyloid
Oligomers

Beta Amyloid
Plaques

Microglia

trials. Unfortunately, there is no evidence that direct interventions with either of these two targets has had any therapeutic benefit, suggesting that both beta-amyloid and tau proteins may be the *result* of the disease rather than the cause.

Consider an analogy.

In diabetes, when the body can't make enough of its own insulin, your cells can no longer use blood sugar effectively, so your blood sugar levels climb. At the same time, because your cells can't get energy from sugar anymore, they begin to burn fats instead. Unfortunately, the result is that your cells produce unwanted acids, which they dump back into the bloodstream. In the middle of the twentieth century, it was considered good medical care to treat the dangerously high levels of blood acid by giving sodium bicarbonate intravenously. This not only didn't help, but actually caused additional complications. The problem was that the elevated acid level in the blood was not a *cause*, but an effect. The proper intervention, we finally realized, was not to treat the high level of acid, but to treat the high level of blood sugar. Once you dealt with the high blood sugar, there was no excess blood acidity to worry about.

In the case of Alzheimer's, we are still working very hard and spending enormous amounts of money to treat the effects rather than the cause, despite the chronic, frustrating failure of these efforts. We are looking "downstream" at the effects, rather than "upstream" at the cause. In reality, Alzheimer's, like many diseases, is not a simple stream of pathology but a cascade of pathologies, yet we continue to focus our efforts at finding a prevention or a cure at the foot of the cascade rather than at the sources of the river.

So the real question is this: What is the source? What happens in the cascade of pathology that results in the disease and, most important, where is the most effective point of intervention?

The actual trigger of microglial damage and microglial cell activation remains unknown. There are hints that various viruses or other microbes may cause infections in these cells, triggering microglial immune responses, resulting in microglial cell division. Because microglial cells are part of the immune system—similar to the macrophages that invade the subendothelial layer of our coronary arteries—this may be a reasonable suggestion. Also, there have been repeated studies that suggest that various antibiotics, such as doxycycline, might help to delay or avert Alzheimer's disease, although none of these studies has shown overwhelming benefits or won general acceptance. In short, while microbial infection is a possibility, we simply don't know why microglia activate and divide.

What we do know is that microglial activation precedes any other obvious pathology, and we know that microglial telomeres shorten prior to beta-amyloid deposition or the presence of tau tangles in the affected neurons. In short, telomere shortening and cell aging precede other changes. While this certainly suggests that microglial cell aging lies "upstream" from both beta-amyloid and tau proteins in the cascade of pathology, one can still argue whether the cell aging is necessary to (i.e., "causes") the pathology or is just one more side effect of the main pathology. While this point of view is reasonable, the logic of the issue strongly supports the suggestion that cell aging is central to the disease—that microglial cell aging "causes" Alzheimer's dementia. All the data line up nicely, and there is a complete lack of contradictory data. Not only does the same basic cascade of pathology occur in other systems, such as in the coronary arteries, but the changes in cell function provide a clear explanation for why beta-amyloid and tau proteins accumulate in the disease.

The most important issue remains that of intervention. Assuming that microglial cell aging initiates the cascade of pathology that results in Alzheimer's dementia, where shall we intervene? We could try to prevent the hypothetical infection, but we don't even know if there is such an infection, let alone how to reliably prevent or cure it. Moreover, once the microglial cells age and become dysfunctional, no amount of antibiotics (even if they were relevant) could stop the pathology. Likewise, once these cells become dysfunctional, it is hard to imagine how we might possibly find a therapeutic agent that would remove the beta-amyloid deposits and dissolve the tau tangles, while still leaving the neurons otherwise healthy and with just enough residual beta-amyloid and tau proteins (in the right compartments) to provide what neurons normally require. No matter where we try to intervene, microglia remain the crossroads for all the damage. The most promising target for curing Alzheimer's is the microglia and the most effective target within the microglia is the telomere, which controls and times the aging of this cell.

ALZHEIMER'S DEMENTIA: QUICK FACTS

Age: Most Alzheimer's disease probably begins in middle age, but is diagnosed only after a decade or two of subtle changes within the cells of the brain. The pathologic avalanche begins long before the first neurons actually die and we begin to see even the most subtle of cognitive problems. Initial clinical diagnosis is usually made after age sixty-five, although early onset forms occur as well. Alzheimer's disease is uniformly fatal. The mean time from diagnosis to death is about seven years.[1] While genetic risks account for a small percentage of cases, particularly familial Alzheimer's disease, the overwhelming risk is age per se. There have been recurrent, wild claims that various agents—for example, aluminum cookware or dietary grains—cause Alzheimer's disease, but few, if any, of these suggestions are supported by data.

Statistics: Estimates on the incidence of Alzheimer's vary. The diagnosis is more frequent in developed nations, due to better health care systems and the fact that more people live long enough to get the disease. But even in developed countries, Alzheimer's statistics are underestimated because the cause of death is often given as only the most immediate cause (e.g., pneumonia). Current estimates are that more than 25 million people have Alzheimer's worldwide, with the incidence climbing along with mean lifespan.

Cost: Alzheimer's is said to be the most expensive age-related disease, largely due to nursing and supportive care. US costs are estimated at more than $100 billion per year and are rising.[2]

[1] http://www.sevencounties.org/poc/view_doc.php?type=doc&id=3249&cn=231
[2] https://www.ncbi.nlm.nih.gov/pubmed/9543467?dopt=Abstract

continued on next page

Genetic risks: The gene most notably associated with Alzheimer's is that for ApoE4, which is one of three forms of an apolipoprotein commonly found in astrocytes and neurons in the brain. ApoE is critical for transporting lipids (such as lipoproteins, fat-soluble vitamins, and cholesterol) and in responding to neuronal injury. While most people have genes for the more innocent ApoE2 (7 percent of the population) or ApoE3 (79 percent) forms of ApoE, those with the gene for ApoE4 (14 percent) are far more likely to get Alzheimer's disease and to get it at a younger age.[3] Those with two ApoE4 genes are at far more risk (ten to thirty times as much risk[4]) of getting Alzheimer's disease than are those without an ApoE4 gene. Nonetheless, the presence of ApoE4 does not automatically result in Alzheimer's dementia and is certainly not known to "cause" it.

Diagnosis: Initial diagnosis is generally based on the patient's or family members' concerns regarding memory loss, intellectual function, or other behavioral changes. Until recently, diagnosis was largely based on clinical examination and neuropsychological testing, but more objective techniques are now coming into clinical use, including laboratory studies of blood or cerebrospinal fluid, radiological studies, or ophthalmological studies of subtle biochemical changes in the lens or retina.

Treatment: At present, there is no way to prevent, cure, or reverse the disease or even to stop or reliably slow its progress. A number of medications have been used (and occasionally still are) that have no measurable benefit, simply because physicians and patients are desperate to try any potential therapy. Such agents include acetylcholinesterase inhibitors, NMDA receptor antagonists, estrogens, monoclonal antibodies to beta-amyloid, omega-3 fatty acids, et al. Several studies have suggested that vitamin E (tocopherols) has slowed the onset of Alzheimer's, although other studies dispute this.

---

[3] http://jama.jamanetwork.com/article.aspx?articleid=418446
[4] http://www.alzdiscovery.org/cognitive-vitality/what-apoe-means-for-your-health

If we want to prevent and cure Alzheimer's disease, the most effective single point of intervention is the microglial telomere. The microglial telomere is the narrow point in the stream of pathology—the point at which we are most likely to prevent the cascade of downstream effects that destroy human lives.

If we want to treat Alzheimer's disease, we want to re-lengthen telomeres.

Parkinson's Disease

As Alzheimer's disease is predominantly a disease of cognitive function, Parkinson's disease is predominantly a disease of motor function. The hallmarks of Parkinson's—altered gait, tremors, rigidity, inability to start or stop walking, "pill rolling" finger motions, language problems—are all grouped together as a problem controlling and coordinating our muscles.

And yet, there are so many similarities between Alzheimer's and Parkinson's that they might almost be seen as one disease manifesting in two different locations within the brain. Whereas Alzheimer's disease attacks the neurons of the cortex, especially in the forebrain, Parkinson's disease attacks the neurons of the midbrain, especially in the substantia nigra and the caudate nucleus. To be sure, this is an oversimplification, as Parkinson's disease can affect wide areas of the brain and the clinical effects of the two diseases—especially dementia and other cognitive changes—overlap substantially. One of the key differences is that instead of seeing deposits of beta-amyloid and tau proteins—as in Alzheimer's disease—we see deposits of alpha-synuclein protein.

Overall, the fundamental similarities between Alzheimer's and Parkinson's are remarkable. In each case, glial cells—especially microglial cells—play a prominent role in the instigation and progress of pathology. In Parkinson's disease, microglial cells and also astrocytes—a glial cell resembling a star—fail early in the pathology. Where tau proteins accumulate and form tau tangles in the neurons of Alzheimer's patients, alpha-synuclein proteins accumulate in the neurons of Parkinson's patients, forming Lewy bodies.

PARKINSON'S DISEASE: QUICK FACTS

Age: While Parkinson's disease is clearly age-related—mean age at onset is sixty—it *can* occur at almost any age. Various risk factors include exposure to any of several pesticides and herbicides, as well as head injuries. It is generally not considered a genetic disease, but genetic predilections do occur and have been defined as resulting from any number of mutations. Nonetheless, the disease's incidence and severity increase with age.

Cost: As usual, costs are difficult to pin down, but the estimate for the US is about $25 billion annually,[1] largely due to the cost of patient care as well as other indirect costs.

Diagnosis: Most Parkinson's patients are initially diagnosed based on history and physical examination. Short of an autopsy, confirmation is difficult, as there is no simple laboratory or radiological test to confirm the diagnosis, although lab tests may be used to rule out alternative diagnoses. Treatment trials are therefore often considered helpful both therapeutically and diagnostically.

Treatment: Because the central feature of the disease is the loss of neurons that produce dopamine, most therapy focuses on drugs such as levodopa and dopamine agonists, which increase dopaminergic effects of those dopamine neurons that still function within the brain. Unfortunately, not only do these drugs have significant side effects, but they lose efficacy as the disease progresses and more of these cells die. In such cases, neurosurgery, brain stimulation, and transplantation of cells (such as stem cells) are increasingly being considered.

---

[1] http://www.pdf.org/en/parkinson_statistics

In both cases, the pathology—as seen in the accumulation of these abnormal proteins—begins in the neurons while the patient is still clinically normal. By the time clinical symptoms occur, the neurons are already dying in wholesale numbers. When the pathology is limited to the midbrain neurons—those in the substantia nigra—the symptoms are largely motor; when the pathology is also seen in the cortex, the symptoms include dementia and resemble Alzheimer's disease. Just as in Alzheimer's disease, the glial dysfunction that occurs in Parkinson's results in failure of the neurons that those glia are meant to support. These cells show failure of intracellular organelles, including mitochondria, ribosomes, proteasomes, and lysosomes. Neurons require healthy glial cells, and when glial cells fail, the neurons are not far behind.

CHAPTER SEVEN

◢◤◢

Slowing Aging

Waiting for Telomerase

Perhaps we will reverse aging tomorrow, but what can we do today?

Help may be on the way in the form of therapies that will re-lengthen telomeres, reset gene expression, and thus halt and reverse aging—therapies that may well be available in the next decade.

But as you read this you may have parents, relatives, or friends who already have age-related diseases, and even if you yourself aren't struggling with such diseases now, the prospect certainly lies in your future. Even if we *can* begin to cure or prevent Alzheimer's disease or heart disease in the next several years, how can we best survive until then? Should we eat differently? Is there a product already on the market that provides protection from disease? What can we do for ourselves and for our loved ones right now?

As a physician, my own concerns are entirely practical, not academic. For me, when it comes to aging disease and even aging itself, the question isn't how does aging work, but *what can we do about it?* I want to know *the single most effective point of intervention.* And until we've proven in human testing what that point is, we all need to know what we can do right now.

A century ago, the same questions were asked regarding polio: If we can't really cure polio, what can we do to prevent it? Just as we worry about the costs of Alzheimer's disease now, a century ago, families worried about the cost of nursing care for their paralyzed children. Until Jonas Salk developed the first effective polio vaccine, there was no protection against the disease. Until we can effectively re-lengthen telomeres, there is no protection against aging. But that doesn't mean that we can't improve our chances of surviving and remaining healthy by engaging in reasonable lifestyle choices. There is no diet or exercise regimen that will stop or reverse aging, but diet and exercise are still the best choices we have if we want to optimize our health—and slow the onset of the diseases of aging—until we are able to re-lengthen our telomeres.

When people ask me what they can do now in order to live longer lives, I tell them that they should be paying more attention to what their doctor or their grandmother tells them, and the grandmother is cheaper. Being human, however, few of us follow medical advice about everyday living, however sensible it may be. If you want to live a long healthy life, eat well, exercise regularly, fasten your seatbelt, and avoid annoying the people around you (you never know who's armed). Unfortunately, people prefer sexier advice, preferably involving miracle foods or startling new forms of exercise. But the plain truth is there's no single food, exercise, supplement, or form of meditation that will stop aging, but there are any number of things you can do to increase your chances of staying healthy longer.

Finally, there is at least one possible product—a telomerase activator—that is currently available that might actually reverse or slow the aging process to some extent while we try to find a more effective intervention.

Caveat: Cui Bono?

There is no shortage of advice about diet and exercise, and most of it is wrong.

There are a few obvious hallmarks of bad advice. For example, the more money it makes for the party giving you the advice, and the more money spent on advertising it, the less likely it is to actually be good for you. Things that are good for us are usually common, inexpensive, and boring. Marketers know that using vinegar on your hair will prevent dandruff, but you can't sell vinegar for a dollar an ounce. If you want to make that kind of money, you have to sell a "specially formulated" shampoo.

The same is generally true of the recurring waves of fad and fashion that drive most human behaviors, particularly when it comes to supplements and diets. It's much easier to hawk products that are *new, improved,* or *revolutionary* than products that were already in common use (even if effective and healthy) a century ago. This is, as any publisher or writer of diet books or cookbooks is well aware, particularly true of fads in food and dietary advice. Marketing a diet is less a matter of whether it works than whether it's new, attractively quirky, and in use by someone who is momentarily famous and attractive.

It's hard to sell oatmeal; it's easy to sell sex appeal.

On the other hand, if you can make a claim that your product has its provenance in ancient history or, better yet, prehistory—a grain eaten by the people of pre-Columbian Peru or a Paleolithic diet—then you can successfully sell your product for a few years. After that, people move on to the next "natural" diet or grain. We tend to believe in the distant past as a guide to wholesome "simpler times," a movement toward what eighteenth-century French philosopher Rousseau called a state of nature, even if that retro-view is false or even dangerous. The distant past is not a reliable guide to any aspect of optimizing your health. If you doubt that, ask yourself this: Did people live longer or shorter lives 100, 1,000, or 10,000 years ago? In lieu of Rousseau, you might better consider Thomas Hobbes's description of human life in a state of nature: ". . . poor, nasty, brutish, and *short*" (my italics). It's like looking at caves as a guide to designing homes: They're picturesque and certainly part of our species' heritage, but scarcely a good way to stay warm, dry, and safe from disease.

The truth is that neither ancient provenance nor the newest "discovery" nor high price nor low is a reliable guarantee of efficacy or credibility. We should be reasonably skeptical of advertising for an expensive product, and equally skeptical of simplistic claims using the words *natural, simple,* or *green.* It's often difficult to discern the truth, and there are few reliable guides other than experience. As in science, so too with advice: If you want the truth, then guesswork is worthless, logic is often good, but data is always best.

One more caveat deserves mention.

For decades, as a practicing physician and a professor of medicine, I have given endless medical advice, yet I have always accepted the fact that not everyone will take it. That's fine with me. My job is not to force people to change their lives; my job is to provide the best advice I can and then let the patients make their own choices. This is true for two reasons. First of all, it's a free country (more or less), and people have the right to make up their own minds, rather than having *anyone* else (including their doctor) make up their minds for them. Second, there is always a possibility that I might be wrong. There is a deplorable tendency for people who are right *most of the time* to assume that they are right *all the time.* They aren't; no one is right all the time.

Anyone's advice should be taken with a grain of salt—especially dietary advice.

I have never demanded that my patients do something, such as quitting smoking. Given advice? Yes. Explained the risks? Yes. Asked if I could help them quit? Certainly. Yet I have never assumed that my profession had the right (let alone the duty) to control the lives and choices of those who come to me for care. My job, one might say, is to make smokers feel guilty. A physician's role is to be an advisor, not a dictator, with regard to your medical care and the path you choose to optimal health. I can suggest a path, offer you a map, and wish you good fortune, but the choices are yours.

This chapter will begin by looking at options that do not directly involve telomeres. Then we'll look at several options that either affect telomeres and aging, or merely make claims to that effect. The chapter will finish with an introduction to what has already been done in the lab, bringing us up to date on avoiding disease and becoming healthy.

The Cliff

Imagine a cliff with a slope in the shape of a parabola: It's fairly level at the top, but becomes gradually steeper until it's almost vertical. We begin our lives well back from the edge of the cliff, young and healthy. Walking is easy, the slope almost level. Then the slope becomes a little steeper, and we begin to see the first onset of aging. As we go further, it becomes harder and harder to stop or even keep our footing. Avoiding disease seems impossible; staying healthy demands much of our time, attention, and effort. Finally, the slope is so steep that we find ourselves tumbling and free-falling, as it were, into disease and death.

Hardly a pleasant metaphor, but it has its purpose.

For one thing, it's useful to think about what happens if we try to live a healthier life. For example, if we are slowly progressing down the slope, what happens if we completely change our diet, and now we eat the absolute best, healthiest, human-optimized diet imaginable? Would this slow our progress, stop our decline, or

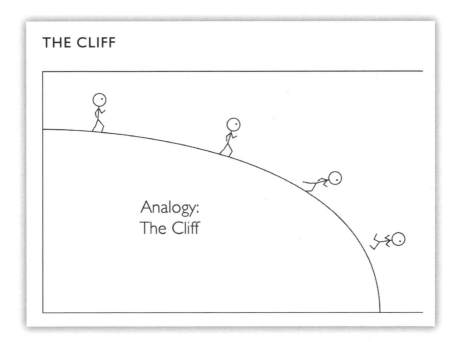

THE CLIFF

Analogy:
The Cliff

might it even help us move back up the slope again? Do we slow, stop, or reverse the progress of aging and disease?

Throughout human history, with no exception, progress down this slope has never been stopped, let alone reversed. From a historical perspective, while we can certainly cure any number of diseases (infections are the best example), we have never done anything that even remotely came close to stopping, let alone reversing the aging process and the diseases that result from aging. On the other hand, there are any number of interventions that can reasonably claim to *slow the progress* of aging.

	Proven Interventions
Speeds aging	UV exposure, smoking, stress, infection, disease, bad genes
Slows aging	Good diet, exercise, immunization, good genes
Stops aging	Nothing
Reverses aging	Nothing

To be more precise, *prior to perhaps 2006*, there has never been a single case of any known intervention that might even remotely be construed as stopping or reversing human aging. We did, however, have a great many interventions that might reasonably be argued to either speed up or slow down the aging process. Specifically, we knew of multiple behaviors or risk factors (such as genes) that accelerated aging, as defined by clinical exam, laboratory values, disease onset and progression, and the mean age of death. Likewise, we knew of multiple behaviors or other factors (such as genes) that appeared to have the opposite effect: slowing the rate of aging and the onset and progress of aging diseases.

None of these—whether it accelerated or decelerated the aging process—was particularly remarkable or obscure. To the contrary, almost without exception, they involved basic, simple, timeless medical advice. To put it bluntly, it was the sort of advice that your grandmother or your family physician almost invariably gave you, and which we ignored completely or at best implemented partially.

To be fair, the advice was often difficult, embarrassing, painful, unpalatable, or time-consuming. It often consisted of advice such as eating vegetables, avoiding calories, sugar, and fats (or anything else that tasted good), as well as getting regular exercise, using sunscreen, washing our hands, and similar items from our culture's store of conventional wisdom and good intentions. We all knew the drill; few of us either followed it or wanted to.

While we cannot change the genes we inherit, there are endless ways—all fairly common—by which we can slow or speed up the aging process and the risk of disease. Within broad limits, there is nothing either original or astonishing about slowing down or speeding up the aging process.

Stopping or reversing aging, however, is an *entirely* different matter.

Despite the panoply of commercial products that claim to stop or reverse aging, there is no example of a product that succeeds at either. There are now foods, supplements, diets, creams, and exercise regimens that claim to stop or reverse aging, but these claims aren't credible. None of these claims—*not a single one*—is truthful. There is nothing on the market that stops or reverses aging.

Well, almost nothing.

It turns out that there are at least two sorts of exceptions to my categorical denunciation of anti-aging products. They are small exceptions, but they're worth looking at. The most important exception is telomerase activators, which I'll discuss later in this chapter. These products speak to the heart of the issue and deserve very careful consideration.

The second exception is that there is some interesting data suggesting that while some medical interventions may not actually reverse aging, they might at least reverse some of the key pathologies seen in age-related diseases, such as atherosclerosis. These interventions do not generally claim to reverse aging, nor are they likely to actually do so, but if they help us avoid age-related *disease*, they are certainly still valuable. These interventions probably reverse some of the obvious changes we see in the process of age-related arterial disease. They may not reset telomere lengths or make our endothelial cells any more capable or functional (or any younger), but if they reverse the disease process and the risk of dying, then that's a notable accomplishment.

The bulk of interventions (and commercial products) that claim to reverse aging, work (if at all) not by actually reversing aging, but by slowing the rate of aging and the progress of disease. Oddly enough, such interventions decrease the rate of aging to about the same rate as people who don't do unhealthy things. In many patients—particularly those who smoke or have high blood pressure, glucose, and cholesterol—the rate of aging is much higher than in those who don't have those risk factors. If we can slow or even stop atherosclerosis and delay mortality, then even if we can't reverse aging, we can improve lives.

For example, if the average nonsmoker has a heart attack at age seventy and the average smoker has a heart attack at age fifty, then we might reasonably hope that if we stop smoking, we might postpone a heart attack by several years, getting closer to the risk of a nonsmoker. If you stop smoking, you haven't actually cured or prevented disease, you've only lowered the risk to the same risks that everyone has. Not a cure, but better than nothing.

With the understanding that most things don't stop aging, and that the most we can usually do is slow the process and lower the risk of disease, let's see what we can do.

Diet

> No disease that can be treated by diet should be treated with any other means.
>
> — Maimonides, twelfth-century physician

Aging cannot be treated by diet, but you can make yourself older and sicker with a poor diet.

First of all, there are no miracle foods (nor any miracle diets, miracle exercises, or miracle meditation) that will keep you young. None. With regard to diets, there are no "good foods" or "bad foods," although I seldom go through a week without hearing one of those two phrases uttered in relation to some food or supplement—for example, "quinoa is good for you" or "sugar is bad for you."

The reality depends on the context.

These days, many people regard sugars as bad, yet the label is far too simplistic. Without sugars, you'd die quickly. Most of your cells rely on sugars from moment to moment as their primary source of fuel, and while some of your cells can use fats or proteins as an alternate source of energy, others can't manage the trick at all. When you eat carbohydrates—pasta, for example—your body breaks down these complex molecules into simple sugars. The question isn't whether or not you need a dietary source of sugars—you do—but how much you need and what form those sugars take. You need sugars to make nucleic acid chains for your DNA and for making various other complex molecules that are essential to life. You need sugars for energy. When broken down into simple sugars, dietary carbohydrates are a useful and readily available source of metabolic energy and are, in fact, the universal "currency" of cellular energy.

There was a time when sugars were regarded as the acme of everything "good" in the human diet. Just before 1900, when both German and American medical physiologists were finally coming to understand the essential role of sugars in cell metabolism, sugar was actually called "nature's perfect food." Some physicians and physiologists went so far as to seriously suggest that the optimal human diet would consist of sugar and nothing else. These days, we recognize that things are a bit more complicated than that. Far from being "nature's perfect food," sugar can create problems. On the other hand, while we laugh at the foolish arrogance of scientists a century ago, are we perhaps making similar mistakes with equal hubris? Is sugar *really* all that bad?

Again, it's a matter of context.

Nature has no perfect food, and where one food is better than another, it is only because it provides a better fit for our cellular needs at that particular time, given our particular genes, with our individual lifestyle. Moreover, many things are fine in the short run, but carry a high risk in the long run. If nature has a perfect *diet*, it's an assortment of many foods, balanced, and appropriate to our individual needs. Life, which is the focus of any dietary discussion, is complex and always more so than we realize.

Nor is there a food *group* that is uniformly bad or good, whether we look at carbohydrates (or simple carbohydrates such as sugars), proteins, fats, vitamins, or minerals. Proteins, for example, provide essential amino acids that we can't build on our own. Without dietary protein, we'd become malnourished and perish of disease. Fats are no different: There are several lipids present in fats that we can't do without. Moreover, fat not only provides an efficient source of calories, but several fat-soluble essential vitamins cannot be absorbed unless the diet contains enough fats to dissolve them. Most flavors are fat-soluble, so meals without some fats tend to be relatively flavorless.

Cholesterol, a lipid, is a good case in point: We have a cultural phobia about cholesterol, yet can't survive without it. Some of us need very little in our diets, some need a good deal more, but cholesterol is not a "bad food" except in context. More broadly, although we are quick to suggest that dietary fats result in arterial disease, and although we know that high levels of lipids and cholesterol in our blood are risk factors, the case against dietary lipids is a lot weaker than most of us are led to think. Lowering dietary fat or cholesterol is no panacea for age-related arterial disease.

What about vitamins and minerals? There is no doubt that we require a good number of vitamins and minerals, although once again, there is more individual variation, probably genetic, than most of us realize. The question, however, is whether or not *supplements* are beneficial. In some cases, they clearly are. No one would argue against supplementation when a person has a vitamin deficiency, but both diets and dietary needs vary between individuals. One person might require twice the vitamin C of another person, or half the niacin. Worse yet, in many developed nations, the typical diet is a poor and unbalanced source of essential nutrients. Just because we stay alive and grow fat on the average fast-food diet, that does *not* mean we are well-nourished. On such a diet, we take in far too many calories, yet may not get all the essential nutrients we need. We can be obese, yet have a profound vitamin or mineral deficiency.

But do we need supplements? Some of us ingest so many supplements that our bodies discard them (especially the water-soluble B vitamins) through our kidneys. A standard joke among pharmacology professors is that Americans have the most expensive urine in the world. Perhaps we do take too many supplements, but what *should* we do?

What's Optimal

More men die of their remedies than of their diseases.

— Molière

The optimal diet is one tailored to *your* genes and *your* behavior. Human beings were not evolved to do well with, nor have they yet adapted to, a high-fat, high-calorie, low-mineral, low-vitamin diet. While some of us might well need vitamin and mineral supplements, what most of us need is a varied diet that includes some protein, many vegetables and fruits, and few simple sugars. That's not a revolutionary suggestion by any means, so let's go a bit further.

Most water-soluble vitamins (the B complex and vitamin C) are fairly delicate molecules and don't like being heated or sitting around for a long time. If your diet consists of cooked vegetables (as opposed to fresh salads and raw vegetables) or, worse yet, no vegetables or fruits, then you are at risk of being low on the water-soluble vitamins and should consider a supplement. The fat-soluble vitamins are interesting in that they depend somewhat on your fat intake: If your diet is too low on fats, again, you may want to consider taking supplemental fat-soluble vitamins (A, D, and E). Vitamin K is an oddball, in that you rely on the bacteria that live in your bowels to make it for you, but it still needs some dietary fat to help you absorb it efficiently. The most common problem, and the diet that really suggests the need for supplementation, is not one that is too *low* on fats, but too *high* on fats. When you get by on fast food and not much else, the "not much else" includes all the vitamins that you're missing. If most of your meals are served through a drive-up window, you probably should have a multi-vitamin supplement for dessert when you get home.

But why not just have everyone take a supplement and be done with it? Ignoring the cost and the bother, which aren't much, there is still no reason to take more than your body needs. Many of us like to believe otherwise; if a vitamin and mineral supplement helps keep me healthy, then surely ten tablets a day will make me live and stay young forever, won't they?

Ah, that it were so.

Unfortunately, there is an optimal amount for each essential vitamin and mineral, and after that things go distinctly downhill. Not only does the data show that additional vitamins seldom help avoid age-related disease, but there is fairly good evidence that the risk of age-related disease and death goes up with unnecessary supplements. More is not only *not* better, it can actually cause disease.

The opposite is also true. We avoid certain things as being "unhealthy," but total avoidance can also cause problems. Just as you can have too much of a good thing, you can also have too little of a bad thing. For example, you can have too *few* free radicals for optimal health. Your body actually needs free radicals to kill bacteria and modulate many of your metabolic processes. So while it's not a good idea to have an overabundance of free radicals in your cells, neither is it healthy to have too few. Also, consider the case of oxygen. Too many oxygen atoms in your cells, particularly free oxygen radicals, can create a lot of damage, but if you don't have *any* oxygen then you die immediately. There is an optimal concentration of everything. For supplements, as well as free radicals and oxygen, moderation may be boring, but it's also your path to health.

How unfair! It would be easier if there really were "good foods" and "bad foods," and the same for supplements, vitamins, minerals, fats, free radicals, and so forth. The reality is complex: None of the things you eat are good or bad except in context and in optimal amounts, and the optimal amounts vary from one person to the next.

So while supplements have their place, that place is a conservative one. Some vitamins may be good, but more vitamins are not better vitamins. Don't let me stop you from taking a daily multivitamin, but don't expect that it will protect you from either aging or age-related disease. It won't. The possible exception has always been the argument about the tocopherols, such as vitamin E. Over the past two decades, there have been suggestions that if you at least double the recommended intake, you might delay the onset of Alzheimer's disease. While the data has never been overwhelming and is frequently argued away (only to rise again a few years later), I have always argued that it wouldn't hurt to try it anyway. After all, the costs and the risks of vitamin E aren't great. High doses inhibit platelets and blood clotting, but don't we recommend aspirin for the same reason? Because we currently have nothing whatsoever to offer patients with Alzheimer's disease, why *not* try a tocopherol supplement? Purists argue against it, but purists are seldom the ones with Alzheimer's, so I leave it to my patients to make up their own minds. It may not help against Alzheimer's, but then, what does?

Specific Diets

When it comes to food in general (as opposed to supplements), we move into territory that is far more familiar to most people. Modern nutritionists have come up with an excellent, simple rule: Eat food found at the perimeter of the supermarket, not the aisles in the middle. If you follow this rule, you will avoid the processed foods, canned foods, sodas, and empty-calorie snacks in favor of fresh vegetables, fruits, meats, and dairy products along the walls.

THE "OFF-THE-WALL" DIET

Never buy food from the aisles in the middle of the supermarket. Buy the food that's found along the outer walls.

Another common suggestion is to never eat anything that your great-grandmother wouldn't have recognized as food. So there go your liter bottles of soda, brightly colored, sugar-laden cereals, and "fruit products" in a plastic tube. Your grandmother's advice, like your doctor's, may not be exciting, but taking it is often particularly effective. A related rule echoes the point made earlier about products' prices versus their value: *Pay the farmer, not the packager.* The higher the cost of advertising, processing, and packaging, the less nutritional value in the product. With some cereal products, you might be better off eating the box than the product inside it.

A similar rule applies to ingredients. The longer the list of ingredients, the lower the likely nutritional value, especially if you can't pronounce the ingredients or have to do a Google search to understand what they are. No one has to go online to figure out what's in a banana.

There's a reason for that.

Buy food, not food products.

In our culture *diet* is a loaded word with two common meanings. One has to do with nutrition, as in, "Do you eat a balanced diet?" The other has to do with our national obsession with losing weight: "I'm going on a diet." That's too bad, because this should be one discussion, not two. A "normal" diet should be relatively low in calories, while having sufficient calories, fats, carbohydrates, proteins, vitamins, and minerals to meet our complex dietary needs. It would be nice to believe that staying on such a "normal" diet would help keep us thin and healthy (which are not necessarily the same thing) and in a general way that's true. It's the specifics that are the problem.

The first problem is that very few of us have such a "normal" diet. Just as common sense is quite uncommon, a normal diet is quite abnormal. Instead, our diets have more to do with rapid preparation and pretty packaging than with nutrition. Yet even if we eat a good, balanced diet, many of us have genes that tend to keep us fat, while others eating and living the same way are thin.

Consider why that might happen.

Our metabolisms have been honed to respond to an environment that's very different from that of the developed world in the twenty-first century. If we could go back a thousand years and visit any small village in Europe, Africa, or Asia, we'd find that few of our ancestors were obese, but the ability to *become* obese—the genes that helped individuals store calories in their body fat—was useful to survival. Such genes helped your ancestors survive a bad harvest and a long winter. Skinny people died of starvation, being unable to store enough calories to make it through to better times, while somewhat fat people made it through until the following spring. Until very recently in human history, being plump was considered a sign of health and prosperity and that wasn't unreasonable.

These days, the opposite is the case. There is generally enough food throughout most of the developed world that, from an evolutionary standpoint, storing fat is no longer a benefit, nor is being unable to store it a disadvantage. And some of us get so fat that we are more at risk of chronic disease, especially arterial disease, diabetes, and joint problems. As the availability of food changes, so too does genetic risk.

You are what you eat, but you are also what your ancestors ate.

Your Diet or My Diet?

> What some call health, if purchased by perpetual anxiety about diet, isn't much better than tedious disease.
>
> — Alexander Pope

The amount of genetic variance is astonishing. If we look at a single gene within a population and compare it to a reference set of actual genetic alleles, we find that the average person has between 5 and 10 million single-nucleotide variants. We are all slightly different from those around us, with slightly different alleles for the same genes. Many of these variants will have no measurable effect on disease, but they may have more subtle effects, including the way we respond to diet. That is, our personal genetic makeup will determine our *optimal* diet and how we respond individually to foods.

In addition to differences in genetics and gene expression, we also differ in our internal bacterial environment, our microbiome. There is growing evidence that this is true of the bacterial microbiome that we carry around in our intestines. Even with equivalent diets and equivalent genes, a different bacterial population can result in one person being healthier, thinner, or happier than another. Food allergies, diet sensitivities, and a host of other problems are probably due more to our bacteria than to our parents, although it's hard to tease out the precise causes. If you really want a healthy diet, you should also be wondering how you can manage to acquire a particularly healthy set of friendly intestinal bacteria. While you might start with the usual approaches—yogurt comes to mind—we have only begun to understand a small glimmer of the importance that probably lies in our bacterial differences, and we still know almost nothing about how to acquire the optimal set of bacterial parasites.

And yet, despite our differences in genes, gene expression, and intestinal bacteria, there are people who suggest that we should all eat a specific "perfect" diet. While there are general truths to specific dietary advice—avoidance of excess calories, for example— doctrinaire diets that suggest a narrow approach and that make no allowance for individual differences are likely to cause disappointment. One example is the currently popular "Paleolithic diet," which attempts to replicate how early humans might have eaten. While the suggestion that we should be eating less in the way of processed food is laudable (*pay the farmer, not the processor*), this diet makes three mistakes. The first is that it ignores individual differences. We are all different (for that matter, so were our Stone Age ancestors) and will react differently to each diet. The second is that our knowledge of what constituted our prehistoric ancestors' diet is probably stunningly inaccurate. Fifteen thousand years is a long time; we scarcely know what most people ate a few thousand years ago, or in the case of some cultures, only a few hundred years ago. The third mistake is in assuming that we haven't evolved since Paleolithic times. To the contrary, our genes respond remarkably rapidly to the foods available within our environment.

Consider the case of lactase, the enzyme that allows us to digest milk. All of us have lactase at birth, enabling us to digest human milk. However, many humans lose the ability to make lactase long before they reach maturity. Paleolithic people almost certainly did not make lactase as adults, nor could they digest milk products. And yet, adult lactase expression—"lactase persistence"—has evolved at least twice in history (and probably dozens of times) in cultures like those of the Masai or northern Europeans, who have adapted quite well to adult diets that include dairy products. We are far from being Paleolithic humans anymore, nor would a Paleolithic diet necessarily suit our genes (or our intestines) these days. While many diets in developed nations are certainly unhealthy, it is unlikely that a presumptive Paleolithic diet (let alone a *true* one) would be optimal for any but a few of us.

Any doctrinaire diet has risks. On the one hand, vegetarians need to ensure they get enough vitamin B_{12}. On the other hand, those who avoid vegetables often don't get enough folic acid. And yet, diets have become wildly successful fads, often going from one extreme ("only eat carbohydrates") to the other ("avoid all carbohydrates") within a few years. Occasionally, two wildly contradictory diets are vying for social acclaim and media attention, even though neither of them is likely to be very healthy. A good rule of thumb is that if a particularly novel diet has become ubiquitous on TV, the Internet, and in bookstores, two things are true: (1) Someone is making a lot of money pushing that diet; and (2) you would be a fool to follow their advice exactly.

> Finding the right diet demands attention to your body and to reality.
> Diet is not a religion, a political philosophy, or a dogma.
> Pay attention to what your own body tells you.

Aging and Diet

Regardless of all the other concerns discussed above, dietary needs shift as you grow older. As telomeres shorten, cells slow down and use less energy. So as you age, you need less food in general. It would be nice if eating more food forced your cells to use the extra nutrition to repair themselves, but it doesn't work that way. One of the most common effects of changing gene expression in aging cells is slower metabolic rate as the cells no longer repair, replace, and recycle as avidly as when they were younger. Therefore, older cells have a lower demand for calories, with the result that excess caloric intake is shunted into body fat.

As we age, we need fewer calories. If we keep eating the same number of calories, we get fatter.

> When you are fifty years old, eating like a twenty-year-old doesn't make you any younger, it just makes you a fatter fifty-year-old.

Most young adults notice that after they finish college, they begin to gain weight, unless they cut back their caloric intake. To a large extent, this particular weight gain results not from a shift in overall metabolic rate, but from a decline in physical activity at this stage in life. As we grow older, however, there is a general overall decline in cellular metabolic activity independent of physical activity. No matter how much tennis you play, your cells simply aren't turning over protein (and other) pools as quickly. While this saves energy, there are two detrimental outcomes. The first and most important is that cells become dysfunctional and begin to accumulate damage, as discussed in Chapter Two. The second detrimental outcome only occurs when we continue to ingest the same number of calories that we did when we were younger, as many of us do. The result is obesity and the increasing risk of the many chronic diseases associated with it.

Optimally, we would adjust our intake of calories and proteins to match our diminishing needs. In general, our nutritional needs decline with age. However, our need for certain specific vitamins and minerals may not decline but remain relatively constant or even increase somewhat. Making recommendations more precise than that of a daily vitamin supplement becomes hard as we grow older, due to our genetic differences and our individual predilections for certain age-related diseases.

DIETARY ADVICE FOR THE AGING

1. Decrease your total calorie intake.
2. Avoid empty calories and choose a variety of foods.
3. Ensure a good intake of vitamins, minerals, etc.

The bottom line, however, is fairly simple. As we get older, we have an even greater need to eat a balanced diet and cut back on our caloric intake to reflect actual metabolic needs. There are very few healthy old people who eat unbalanced, high-calorie diets.

Eat your age.

Exercise

Walking is the best possible exercise. Habituate yourself to walk very far.

— Thomas Jefferson

The conventional wisdom is that exercise is good for you and that it delays aging and disease, but is any of that actually true? Probably to an extent, but less than you might think. There is certainly no evidence that exercise slows aging, although it may well help you avoid age-related diseases. You should definitely value exercise, but with some sense of rationality.

For one thing, if exercise is so good for you, then why do the ads often tell you to "check with your doctor before beginning our exercise program?" Some people not only won't be improved by exercise, but might not survive taking it up. On the other hand, it's not so much the exercise as it is the underlying health problems that constitute the risk. The classic example is the elderly man who, frightened by new chest pains, goes to see his physician and is told he has angina and is at risk of having a heart attack. It's certainly likely that his condition might improve somewhat if he gradually eases into a reasonable exercise program—but if he goes out the next day and tries to run ten miles, we'll hardly be surprised if he has a heart attack during his run. Exercise is good for you, but only in context. If you want to exercise your body, you also need to exercise your brain and not do anything stupid.

A second caveat is that a great deal (though not all) of the data suggesting that exercise is good for you is correlational and not causal. If I poll 2,000 people and find that 1,000 of them are exercising every day and have never had an illness in their lives, while the other 1,000 never exercise and are uniformly unhealthy, it doesn't prove that exercise is good. The group that is exercising might include only teenage athletes, while the group that doesn't exercise might include only elderly patients in hospice care. Teenagers tend to be healthier than elderly hospice patients whether they exercise or not.

Although this example may sound silly, many studies of the "benefits of exercise" make essentially the same mistake. Assume for example that we have 2,000 people, all the same age and all with the same medical history, but half are known to exercise and half don't. It may be that the half that don't exercise have any number of "high-risk genes" that not only will eventually lead to chronic disease, but that also result in low energy levels; these people don't feel like exercising in the first place. In this case, disease is due not to a lack of exercise, but to the genes that result in both disease and a sedentary lifestyle. In short, some people are born with none of the luck.

The critical question is: What happens if, as an experiment, we take a group of people who aren't all that healthy and who have known risks for disease and get them to exercise regularly? Does this make them healthier? Does it prevent disease? While that sort of study is remarkably hard to do, we still have some idea of the answer.

Not surprisingly (and assuming you don't do anything particularly stupid), exercise really is good for you, for many reasons. For one thing, it tends to lower blood pressure and serum glucose levels, thereby decreasing your risk of any number of chronic diseases. For another thing, depending on the type of exercise you do, you can slow the onset of osteoporosis by creating a recurrent load across whichever bones are involved. Running, for example, helps retain bone density in the legs, while doing "dismounts" in gymnastics can help retain bone density in the vertebrae. Roughly speaking, it's a case of "use it or lose it." Whatever you tend to do a lot of will tend to keep that part of your body in good health.

One of the benefits of exercise may surprise you—the value of motion for the cells that line our joints. Essentially, our joints do best when they undergo recurrent gentle motions under some joint tension (or gravity). The reason is that the chondrocytes within our joints have no direct blood supply and they rely on motion to stay healthy. All nutrients and oxygen must diffuse in from relatively distant capillaries, and waste products and carbon dioxide must diffuse outward the same way. It's like trying to get black ink out of a kitchen sponge using tap water. Each time you squeeze the sponge, some of the ink runs out; release, and it absorbs clean water. If you repeat the process, then eventually the sponge becomes clean. Essentially the same thing is happening to our chondrocytes: Alternate compression and relaxation across the joint surface serves to exchange both nutrients and waste products, as well as gases. Joints that are in more or less continual, easy use every day will be better off than joints that are never in use. In general, joints—like muscles, arteries, the heart, and the lungs—do better with exercise.

Unfortunately, as you might expect, there are exceptions.

Let's say you're twenty years old. You jump high and land hard on your knees. You crush a few joint cells—chondrocytes—but the remaining cells divide and replace the damaged ones. Of course, that ensures that those cells have shorter telomeres, so if this violent jumping and landing is a habit of yours, they age a bit faster than they would if you spent your time walking instead of jumping around. This is why professional skiers and basketball players tend to have "old" knees—early osteoarthritis—and why they need knee replacements at a relatively young age. Here, it's not a question of nutrients for the joint tissue, but of actual injury. Likewise, if you are a carpenter or a stonemason, and you make a habit of hitting your knuckles with a hammer, you can expect osteoarthritis to develop early in those joints. Exercise is one thing, but injury is another.

Anytime your body has to replace cells that you've managed to kill, your body is accelerating its rate of aging.

So while exercise may have multiple benefits and may even delay age-related disease, those benefits can be offset by exercise that damages your body. The human body is made to be used, and no matter what you do, you will age over time, but if you injure yourself repeatedly, you will age a lot faster.

Also, the benefits of exercise do not increase with the amount of money you spend on exercise. Benefits do not accrue from high-fashion tights, $200 running shoes, or joining the most fashionable health club. Try observing how people get from the lobby to the second floor of a public building. Some people take the stair steps two at a time, some drag themselves up the stairs, and some take the elevator. Those in the first group are getting exercise for free. No dues, no equipment. The same is true even of the way we walk. People who walk energetically are constantly exercising. Those who take shortcuts and move slowly get far less benefit from their daily movements. Just because you're upright doesn't mean you aren't "sedentary."

Effective exercise may consist of simply using the stairs at work—or gardening, dancing, or even walking around the house *energetically*. The more you think of exercise as requiring a club membership, a special time of day, a specific set of clothes, or as a dismal way to spend an hour, the less likely you are to obtain benefit.

Exercise is what you *do,* not what you *spend.*

EXERCISE ADVICE FOR THE AGING

1. Move your *entire* body: every joint, muscle, and bone.
2. Stretch everything: Joints were meant to be moved and used.
3. Have a daily basic exercise, then add variety.
4. Try to add exercise to your day by using the stairs instead of the escalator, walking instead of driving; in other words, keep moving.
5. A few minutes every day is better than a few hours on a weekend.

Meditation

The aware do not die. The unaware are as though dead already.

— The Dhammapada (sayings of the Buddha)

What about the value of meditation?

If you spend as much time among the anti-aging community as I do, you certainly hear a great deal about its supposed benefits.

It depends what you hope to accomplish. Meditation might not take away years, but it certainly might make you more aware of the years you have. Most longtime meditators swear by its benefits, and why not? They wouldn't do it unless they felt they got something from it. There is, therefore, little doubt that meditation has *subjective* benefits, but the issue of measurable, *objective* benefits is less cut-and-dried.

There are innumerable forms of meditation found within numerous religious and cultural settings. Those who study meditation scientifically tend to divide meditation into two basic types on the basis of both the instructions given to neophytes and the outcome as judged by EEG (electroencephalogram) changes. During meditation, the response to external stimuli can be markedly distinct from the normal, non-meditative state. Essentially, one type of meditative practice (for example, Zen) results in a continual awareness of external stimuli, yet the meditator doesn't habituate to those stimuli over time. The other type of meditative practice (for example, yoga) results in little or no response to external stimuli, let alone habituation. Compare these two types of meditation to normal cognitive states, in which most of us respond quite reliably to external stimuli, but then rapidly grow used to repetitive stimulation and stop responding after time. We hear the clock ticking at first, but then stop paying any attention (and our EEG stops responding) as it continues. The Zen meditator might continue to hear the ticking; the yogic meditator might ignore the ticking from the outset. In either case, meditation does change the way we attend to our environment, at least during the meditation itself.

Does this help us in any way when we are no longer meditating?

When asked about the subjective value of meditation, many cite it as helping to "lower stress" or "become less angry or emotional." For many others, the benefits are even more positive, often described as resetting or centering. Meditation resets your mind, so that instead of flying in a thousand directions, you focus on the task at hand. Or meditation puts you back in a center, much as clay on a potter's wheel first needs to be centered because otherwise it flies off the wheel altogether. Most meditators see no particular need to prove that it "works" any more than they need to prove that they enjoy gardening or cooking. Some activities are simply enjoyable. We aren't required to rationalize them. However the benefits may be described, it is clear that among those who meditate there is a strong belief in a subjective benefit. Objective benefits, however, might be another matter.

Which brings us to the question of whether meditation can help extend our healthy lifespan.

Hundreds of studies have attempted to evaluate potential physical benefits of meditation. A good many of them start with an agenda, continue with poor experimental technique, and derive precisely the results that the researchers want to find. Many other studies have been done carefully and conscientiously with the goal of actually finding out if there are benefits or not. It's not easy to tease out the facts, but it's clear that while there are measurable benefits, there are no panaceas, particularly for aging and age-related diseases. Most of the benefits involve measures of physiological stress, such as blood pressure and immune function, and some of these are correlational, not causal. So if we want to know whether or not meditation can, for example, delay Alzheimer's, the answer is still arguable (and argued).

What about telomeres? Specifically, can meditation extend telomeres or can it slow down telomere loss? This is the modern, quantifiable version of the question "Can meditation make you younger?" Several studies have looked at telomere lengths over time and have suggested that meditation can extend telomeres. Unfortunately, the data doesn't actually support the conclusion. One problem lies in using telomere measurements in peripheral white blood cells as a reliable measure of the body's overall aging

---

MEDITATION ADVICE FOR THE AGING

1. Meditation is an invitation to lower stress.
2. The form is less important than the quiet.
3. Find a daily time and place and keep it.
4. Two minutes a day is better than an hour once a month.

---

status. Longer telomeres in your bloodstream may imply that you are under less stress (for example, from an infection), but they don't mean that the telomeres in your bone marrow, let alone the rest of your body, are any longer or that you are any younger.

Meditation's value is not how much life you will have tomorrow, but how much life you have today.

Telomerase Activators

No form of exercise, no matter how strenuous, no diet, no matter how remarkable, and no form of meditation, no matter how profound, will prevent aging. While it's true that we can accelerate aging, there is no behavioral or dietary change that can stop or reverse it.

But there *are* ways to slow and reverse aging.

We have known for years that we can reverse aging in cells, tissues, and, more recently, in animals. The question that remains is how *effectively* can we reverse aging in humans? The field is changing rapidly, and the public is slowly coming to understand the possibility of resetting gene expression by resetting telomere length. At the moment, there are dozens of products being hyped, most without any justification, all claiming that they are effective at re-lengthening telomeres.

Geron Corporation identified the first of these effective activators more than a decade ago and licensed it to TA Sciences shortly thereafter. These activators, which were discussed in Chapter Four, are based on astragalosides and have been shown to be clinically effective in reversing certain aspects of aging. Specifically, there are two published studies of people who have taken TA-65, one looking at immune function and the second looking at several other biomarkers of health and aging. In both studies,[1] there was evidence that telomere lengths were affected in most patients, and in both studies there was evidence of "rejuvenation." Although not everyone had the same effects, a number of patients had immune function improvement equivalent to about a decade in age (including fewer senescent T cells). Similar results were found for blood pressure, cholesterol, LDL, glucose levels, insulin levels, bone density, and other measures felt to reflect age-related disease.

For those wanting to take an active compound that might effectively slow or reverse aging, the use of telomerase activators is tempting, but several caveats deserve mention. The first is that these two studies were small, and the results were neither overwhelming nor inarguable. For example, the immune changes were seen predominantly in certain patients who had already had a history of cytomegalovirus infection. The second caveat is that the changes seen are only in biomarkers rather than an actual disease. For example, lowering your cholesterol is probably beneficial, but actual improvement in your coronary arteries would be better, and reducing the rate of heart attacks would be best. Biomarkers like cholesterol, however good they may be, are not diseases, and they don't themselves cause death or aging. A third caveat is that despite individual claims and despite the data, there is no evidence that anyone on TA-65 or any other telomerase activator ever got any younger.

No one went from age seventy to age forty; it simply didn't happen.

Realistically, we could easily argue that TA-65 and perhaps other telomerase activators might reverse some aspects of aging, but my best estimate is that such compounds are only about 5

---

[1] Harley, C. B. et al. "A Natural Product Telomerase Activator as Part of a Health Maintenance Program." *Rejuvenation Research* 14 (2011): 45–56. Harley, C. B. et al. "A Natural Product Telomerase Activator as Part of a Health Maintenance Program: Metabolic and Cardiovascular Response." *Rejuvenation Research* 16 (2013): 386–95.

percent as effective as they need to be if we want, for example, to cure or prevent Alzheimer's disease. The data is suggestive and intriguing, and we certainly might consider taking a telomerase activator even now, but we need a much more effective intervention if we want to cure and prevent aging and age-related diseases.

From a practical perspective, should we take a telomerase activator?

At the moment, the cost is high—several hundred dollars per month—and the evidence is suggestive but not overwhelming. Worse yet, while any number of companies are offering cheaper versions of these products, there is no certainty that any of them actually contain active astragaloside compounds or, if they do, that they contain the most effective ones. Moreover, there are competing claims that other compounds—such as resveratrol, TAM-818, et al.—work as well as or better than astragaloside compounds, but there is little or no data to support these commercial claims.

In short, there is at least one product with a scientific basis for making claims that it affects aging, along with dozens of other products that may be cheaper, may or may not work, and have no data to support their use. With most of these products, there is no rational basis for why they *should* work. Others might work, but there is no data suggesting that they do. A small number probably work, but are poorly understood so far, and their safety, legality, and clinical effects are arguable.

What is clear is that telomerase activators are neither a commercial gimmick nor snake oil. Although the market is fraught with false and unsubstantiated claims, telomerase activators have been shown to be effective in cells, tissues, animals, and—to a limited extent—in human beings. As of this writing, there has not been

---

SHOULD YOU TAKE A TELOMERASE ACTIVATOR?

Yes, *probably*.

The decision is a bet, based on cost, data, and your finances. TA-65 has supporting data and may have health benefits.

any evidence of significant side effects or risks, such as cancer. The major questions are:

1. Which specific commercial sources are effective?
2. How effective are the available telomerase activators?
3. Is the cost worthwhile, given the individual's budget?

We know telomerase activators work to reverse aging and that nothing else does.

New technology drives the ability to gather new data, but new data should drive new understandings. Just as Leeuwenhoek's microscope let him see "animalcules," which drove a new understanding of human disease, so did Hayflick's careful experiments drive a new understanding of cell aging, and the experiments on telomerase activation are now driving a new understanding of aging in general.

CHAPTER EIGHT

⟋⟋⟋

Reversing Aging

The Potential

Within the next decade, we will more than double the healthy human lifespan.

We are at a pivotal point in human history—one that will be seen as such hundreds or even thousands of years from now. We now have the knowledge and the ability to intervene in aging and its diseases.

At the turn of the twenty-first century, we showed for the first time that we can reverse aging in human cells and human tissues. Over the next decade, we began the first trials of oral agents that promised to at least partially reset the aging process at the clinical level in human beings. Beyond that, a number of academic laboratories—including those of Maria Blasco in Madrid and Ron DePinho at Harvard—had shown the ability to reset aging in animals, using a variety of different methods. In every case, without exception, resetting aging has been accomplished by re-lengthening the telomeres, thereby resetting the pattern of gene expression, resulting in healthier and younger function not only in tissues, but for the entire organism.

We are on the brink of an enormous leap forward, in which we will become capable of reversing the aging process in an obvious and striking way. We are about to not only cure and prevent age-related diseases, but reset the aging process itself.

A hundred years from now, what will be the date that school children learn as "year one" of a new era of human health and longevity? It could be 1999, when we first reversed aging in cells, or 2007, when the first oral telomerase activator became available. Perhaps it will be sometime in the next few years, when we begin human trials of telomerase to cure Alzheimer's disease. In any case, the date will be within the lifetime of most people living today. Reversing human aging, curing age-related disease, is imminent. Posterity will point to a specific year in these first two decades of the twenty-first century as the moment of the single most important advance in medical history—a moment that changed forever what humans are.

Strangely, few people are aware of the changes that have already been underway for the past two decades. Often, the progress has been hidden among smaller, more prosaic advances. A number of large research foundations have been funded during those twenty years, earnestly devoted to understanding aging and age-related disease, yet almost without exception, these have continued to work within old paradigms, and thus have produced little clinical benefit for the aged. None of these research foundations, despite their funding and strenuous effort, has been able to change the clinical outcomes of aging. This is not the only example in medical history of funding based on current paradigms rather than future advances.

In the early 1950s, prior to the first polio vaccine, similar investments were made in improving the iron lung, better nursing care, and futile clinical treatments for children with paralytic poliomyelitis—electricity, oxygen, herbs, and high doses of vitamin C among them. People hoped for a cure, yet we put enormous amounts of money and effort into interventions that were neither fundamental nor effective. We have done much the same for aging and age-related diseases. With the assumption that aging is inevitable, we treat the complications and symptoms while ignoring the causes. Many still cling to wear and tear, free radicals, "aging genes," and other simple paradigms that offer no clinical return on the large financial investments made in aging research.

There is nothing so terrible as activity without insight.

— Goethe

The key advances in aging have been made almost incidentally, and certainly without fanfare, by a few small biotechnology firms, a few bright researchers, and a few insightful clinicians. Those working at Geron, Sierra Sciences, and TA Sciences have contributed, as have several academic researchers. Even my own books and articles have been among the critical pieces. Certain people have looked a bit more carefully, thought a bit more deeply, and worked a bit harder, allowing us to make advances that would otherwise never have happened. That is what happens when paradigms shift and scientific insight takes us in unexpected directions.

When we talk about putting an end to aging, people often leap to wrong conclusions. So before discussing what we are about to accomplish and what it will mean to each of us, let's be clear about what will *not* happen. First, we will not achieve immortality. That is the stuff of myth, fantasy, and science fiction. No matter how healthy, no matter what your genes or your gene expression, life will remain limited by—if nothing else—violence, accidents, acute diseases, and simple misfortune.

Second, when we talk about radically extending the human lifespan, the first reaction of many people is, "Why would I want to live that long?" That is, why would one want to spend a century or more in a nursing home? Of course, that's not what I'm talking about. The mistaken assumption is easy to understand, because increases in the *average* lifespan in developed nations have come partly because we are able to keep old, sick people alive longer. Our misconceptions and the fears they engender are stoked by fiction and mythology. In Greek myth, Eos, immortal titan of the dawn, asks Zeus to grant immortality to her human lover Tithonus, but neglects to have eternal youth written into the contract. In a kind of horrific practical joke, Tithonus is doomed to a life of eternal decrepitude. In *Gulliver's Travels,* Jonathan Swift gives us the Struldbruggs, immortals who become aged and feeble in mind as well as body. At age eighty, they are declared legally dead, their estates are passed on to their heirs, and they are forced to live meagerly on the dole. And then there is Oscar Wilde's Dorian Gray, who maintains a gloss of youth while rotting away inside.

None of these fearful fictions has anything to do with the reality of reversing aging. Doubling the human lifespan—which is entirely feasible—can only be done by ensuring that we live in good health. We cannot double lifespan if that entails a doubling of Alzheimer's disease, atherosclerosis, and other age-related diseases. There was a time when polio left thousands of children trapped in iron lungs. The polio vaccine did not extend that imprisonment; instead, it gave children the gift of normal childhoods. As we look ahead to the end of aging, the prospect is the same; we will not extend the years spent in nursing homes, but offer the gift of healthy lives. The only way to increase lifespan is to cure and prevent the diseases we fear the most, diseases that put us into nursing homes, where life becomes a shadow of the past.

We can offer health and life, rather than the dimming of the light.

As we extend lifespan by actually reversing aging, we will drastically cut the costs of medical care, erasing the need for nursing homes, preventing age-related diseases, and leaving people healthy, whole, and fully capable of living their lives completely. We have done almost everything we can to increase lifespan for the aging—often at a steep cost, both financially and emotionally—but now, the only way to further increase human lifespan is not to prolong disabilities, but to improve health.

If we reverse aging, if we can prevent age-related diseases, then how long will we actually live? That is difficult to predict. We won't really know until long after we begin to prevent aging—until people have lived as long as they will, whatever that might be. Given what we do know of human biology and clinical medicine, and from the little information we have from animal models and tissue experiments, we can make a guess. Within the next decade or two, the projected mean human lifespan may very well move into the range of several centuries, with far better control of diseases such as cancer, Alzheimer's, and atherosclerosis. We are about to change human medicine—as well as our lives and our society—forever.

> We can at least double the human lifespan and will likely extend the average lifespan to several centuries of active, healthy life.

The notion of healthy human lifespans in the range of, say, 500 years is an entirely rational point for argument, even in terms of what we only now understand about possible interventions in the aging process. Once we can extend the healthy lifespan by several hundred years, it will, obviously, take us that long to know how well we have actually succeeded.

In short, this will be the longest experiment in the history of science. We'll all have to wait and see.

The Pathways

There are four pathways by which we can reverse aging, and three of them are already being actively explored. The most elegant solution will be to use telomerase activators, drugs that "turn on" our own telomerase (using the hTERT gene) and thereby reset gene expression. This first pathway is being actively developed and tested by a number of biotech firms (for example, Sierra Sciences), researchers, and academic laboratories worldwide. So far, there are at least two potentially effective agents on the market, although it's not clear how effective they actually are, and neither appears to be as effective as we would wish. What data there is suggests that the astragaloside compounds, particularly astregenol, have significant benefits when measured by biomarkers, such as cholesterol levels, that serve as indirect markers for disease. As yet, however, there is no data showing that these compounds directly affect age-related diseases and thereby decrease morbidity or mortality. Also, there is no data on how much these agents might extend the healthy lifespan, if at all.

The second solution is to use telomerase protein. The challenge with this approach is in getting the protein into cells effectively. Until a few years ago, getting a therapeutic protein to enter a cell was not considered feasible, yet a number of researchers have shown that this approach also has potential. One biotech firm, Phoenix Biomolecular, was created in 2005 to attempt this approach, but the company failed without ever beginning clinical trials. No projects are currently underway to test this pathway.

The third solution is to use the messenger RNA for telomerase, a feat first accomplished in early 2015 by Helen Blau's group at Stanford and not yet expanded to animal or clinical trials. This approach has been considered difficult due to the fragile nature of mRNA molecules, making the method feasible for laboratory studies on cells (in vitro), but perhaps too demanding for clinical trial on human patients (in vivo). Whether this problem can ever be overcome remains to be seen, but the approach still remains enticing.

The fourth solution is to deliver the telomerase gene itself (either via liposomes or viral vectors) to the body's cells. There are several groups (for example, Teloctye's use of adeno-associated viral delivery) actively pursuing this pathway, and clinical results can be expected within the next year or so. In both cases, the key is to give the delivery system a proper "address" so that it can enter into the appropriate cells. While this is sufficient to get the telomerase gene into most cells, certain tissues present additional obstacles. The brain, for example, has a blood-brain barrier that restricts delivery. Both obstacles—getting into the right cells and the blood-brain barrier—have already been overcome in animal studies, suggesting that human trials cannot be far behind.

Any of these pathways can be used treat specific age-related diseases. Some groups clearly intend to go for what they see as low-hanging fruit, believing that preventing skin aging and other cosmetic issues will be easier and perhaps more lucrative than curing disease. Some of us, however, are deeply convinced of the promise and clearly see the greater human need, and we're now aiming directly at diseases that have no available treatment, such as Alzheimer's. Knowing we may have the ability to intervene in aging itself, we want to do what no one has done before and what most needs doing. Human telomerase trials to treat Alzheimer's are already planned as I write.

The Medical Outcomes

What diseases will we cure?

Alzheimer's disease, atherosclerosis, osteoporosis, osteoarthritis, skin aging, immune aging, and most other age-related diseases will recede into human history rather than menace our personal futures. Some exceptions will remain. While we can undercut the most important cause of strokes, which is associated with aging, strokes will still occur due to trauma or genetic predilection, independent of aging changes. Some lung diseases, perhaps including COPD, may still occur due to environmental exposures or toxic damage, as will some genetically induced pulmonary diseases. The precise line between what can be prevented and what can't is defined by the difference between gene-related disease and epigenetic-related disease. If you have sickle cell disease due to an abnormal allele, telomerase has nothing to offer you. On the other hand, where disease is related to aging—with its subtle but pervasive changes in the patterns of gene expression—then telomerase has a great deal to offer each of us.

Whatever the approach, direct genetic delivery or telomerase activator, telomerase therapy promises to eradicate most age-related human disease. Moreover, these are the diseases that have until now been all but impossible to treat. Telomerase therapy will not only be more effective than any other approach to date, but will intervene most effectively in those diseases that have had the least benefit from medical care.

Telomerase therapy will be effective in curing or preventing Alzheimer's and other age-related neurological diseases, atherosclerosis and other age-related vascular diseases, and a host of diseases—such as osteoporosis and osteoarthritis—that are less likely to be fatal yet have high morbidities and have been impossible to arrest or reverse until now. Moreover, most cancer will be prevented as we use telomerase to stabilize the genome, improve DNA repair, and prevent most of the accumulated mutations that underlie clinical malignancy. Telomerase therapy will also work well in treating aging problems in skin, the immune system, and most other body systems.

There will be limits. Telomerase therapy will lower the risks of strokes but not eradicate them, because not all strokes are due to aging. Nor will telomerase reverse problems that have progressed far beyond your body's ability to repair. It's impossible to repair cells or tissues that no longer exist, such as joints that have already been surgically replaced, long-dead neurons (as in chronic Alzheimer's disease), or dead muscle tissue after a major heart attack. These are "Humpty Dumpty" problems; neither all the king's horses nor all the king's men nor any telomerase therapy, can ever put Humpty together again. Telomerase therapy can't fix what no longer exists.

Nor can telomerase therapy help with the hundreds of diseases caused by specific gene problems, such as sickle cell disease. Telomerase therapy can optimize the pattern of gene expression, but it can't replace the genes. All of us were born with genes that limit our bodies. While telomerase can't redefine those limits, it can prevent disease within them. Think of genes as a set of tools that you inherit; while telomerase can't change the tools, it can ensure that you use them capably and efficiently. Equally, telomerase can't cure high-risk or self-destructive behavioral problems. If you overeat, smoke, drink or take drugs excessively, or drive race cars for a living, then you might as well skip telomerase therapy, because you may not live long enough to need it.

Even given its limits, however, telomerase therapy promises to give us an entirely new approach to our most common diseases, allowing us to treat—effectively and cheaply—diseases that until now have either been ignored ("That's not a disease, that's just old age") or have been unresponsive to any therapy.

What Will Telomerase Therapy Be Like?

Treatment is surprisingly simple. A day in the clinic might go something like this:

You arrive at your physician's office much as for any other medical visit and are shown into a treatment room. The nurse starts an IV and you see a small, clear-plastic bag of translucent fluid—much like any other intravenous medication—attached to the IV. It begins to flow into your vein. After half an hour or so, the nurse rechecks your vital signs, and a few minutes later you are on your way home.

Perhaps two weeks later, you return for a second treatment identical to the first. A few weeks after that, your physician checks your blood tests and confirms that the telomeres in your blood cells have been reset. Depending on your prior medical history, your physician checks other laboratory values, runs some cardiac tests, or does an MRI, perhaps looking at your knee joints. In every case, subtle improvements become evident.

It takes decades to grow old, and while repair will begin quickly, it takes weeks and months to notice improvements. They begin subtly, within your cells, progress a bit more obviously to your tissues, and eventually become undeniable in your daily life. You begin to notice more energy and a feeling of well-being. Once chronically tired, you find yourself thinking about activities you haven't done in years. Your sleep improves. You wake up without pain. Your memory is back to normal. Your breathing is easier. You are regaining something that you once took for entirely granted: your health.

Welcome to a far longer and much healthier life.

Now to address some other common questions:

Is telomerase therapy a one-time treatment?

You will need the therapy once every decade or so.

How long will it take?

The entire process will take from a few months to a few years to become complete. While the final result depends on how much damage has occurred, the rate of recovery will be similar for most people. If you had early Alzheimer's, for example, you will do much better than someone with advanced Alzheimer's, so while the rate of improvement would be the same for both (over a few months), the final result will be better if the disease hasn't gone too far.

How young will it make me?

The therapy could reset your physical age by several decades, but it cannot make you a child. There is one set of cellular mechanisms that result in adult maturation, but it's the telomeres that control the aging process.

What about side effects?

Because your body will be rebuilding cells and tissues, as well as repairing your cells internally, it will need more energy than it does otherwise. You can expect to have a greater appetite and to experience initial fatigue as your body focuses on healing and repair.

Will I be able to afford it?

The cost of telomerase therapy will be low, due to the huge patient base over which the research and production costs can be amortized. The majority of the cost of telomerase therapy will be due not to research or production, but to distribution and delivery—for example, the costs of starting an IV in a health care facility. These latter costs include hospital overhead, insurance, health care workers, and a host of other "delivery" costs that have little to do with the cost of the therapy per se. The estimated cost of the telomerase therapy is likely to be within the same order of magnitude currently seen for vaccination drugs—less than $100 per patient dose. Even using a pessimistic projection, the costs of telomerase therapy will be extremely low, particularly when compared with the costs of the diseases that telomerase therapy will prevent and cure.

Therapy will be inexpensive, and in fact profitable both to individuals and to society.

The Social Outcomes

All that is valuable in human society depends upon the opportunity for development accorded the individual.

— Albert Einstein

For the individual, telomerase therapy promises health, optimism, and a new way of looking at life. It removes the fear of the future, the feeling that we are approaching a cliff of unavoidable disease and disability, the knowledge that we may suffer the loss of our independence, our health, our loved ones, or—as with Alzheimer's disease—the loss of our own souls.

Do you want to see the world? Learn a language? Have more time to fulfill a dream? You'll have the health—and the time—to do the things that you have long meant to do. But just as when you were younger, you'll still need to make a living. Currently most of us can work into our sixties and then live on our savings, because life expectancy doesn't extend too many years beyond retirement. With telomerase therapy, however, you might live to 200. Most of us can't retire for 150 years on what we saved from forty years of work. But you'll be healthy and capable of working much longer. Jobs may change as well. Why should you spend another half-century on a job you no longer enjoy, when you can take your savings and invest it in learning and starting a new career? You'd live long enough to take that chance—to find a career and a life that you come to realize you prefer. You might even intersperse several careers with several retirements.

Families will also change, as familial bonds will extend over four or five living generations. We love our children, we cherish our grandchildren, but how much time will we want to—or be able to—spend with great-great grandchildren, several generations younger than we are? The attitude of the young toward the old may change when old people are no longer decrepit. Will we grow wiser with many more decades of experience? That's hard to say, but we will certainly have more knowledge and experience to share with the young. As long as humans have existed, the pattern of our lives—maturation, marriage, children, old age—has been a constant motif. But that theme will change and with it all of the social underpinnings of our lives. How many marriages will survive for two centuries? How will families adjust to changing times? How will our children develop new social customs to meet old human needs?

How will larger social structures, even nations, adapt as our lives extend? We will have the benefit of greater long-term interest in peace and prosperity, but there is also the prospect of conflict arising from inequality if extended lifespans are only offered to some and not to all. Telomerase therapy promises so much at so little cost that its potential to expand medical inequality is small, yet the possibility is fraught with risk. War is historically planned by the old and fought by the young, and both groups will see changes that may affect the likelihood of international conflict. How that likelihood changes is unpredictable, yet critical to our future. There is also the chance that living longer will prolong what might otherwise be single-generational problems. In the decades after wars, terrorist acts and crimes, as well as anger and the desire for revenge, fade with passing generations. As people begin to live much longer, will such hatreds live on and affect national policy in ways that perpetuate conflict?

There are more optimistic notes as well. We may well place greater value in protecting our long-term economy, our global environment, and our children's interests as we realize that we will still be alive to face the outcome of our choices. No longer would we be able to ignore the long-term ramifications of our political decisions. National liabilities that previously extended for two or three generations would persist for one generation. We'd have to live with and be held accountable for national debts and obligations. While we might be more motivated to avoid debt, we might also be more likely to invest in the future—in science, education, and exploration of many kinds. Long-term plans and investments for such things as space elevators, asteroid mining, or cities on the moon would have more immediate implications.

Population can be expected to rise. Roughly speaking, it would climb as fast as the average lifespan climbs, at a time when we already have a keen fear of global overpopulation. On the other hand, not only is population growth markedly decelerating in almost all countries, but UN projections have caused many to begin to fear (and plan for) *falling* population densities within the next century. This is happening even now in some countries.

A growing economic problem in many developed countries has been a reduction in young people supporting growing numbers of older and infirm adults. Telomerase therapy would undercut the basis of this problem, since older people would no longer be infirm and thus would be quite capable of caring for themselves. Telomerase therapy promises an unprecedented demographic revolution, in which increasing numbers of healthy adults actively participate in and benefit the economy, and offer a wealth of experience and knowledge. We'd no longer bury this store of knowledge, but actively use it to improve our economies and our societies. And this store of knowledge would be far greater in each individual. Today, the time one devotes to higher education and on-the-job or "lifetime" learning might span forty or fifty years—sixty or seventy if we include retirement. Imagine what might happen if that period grew to 150 or 200 years. The old will be repositories of vast knowledge that continues to be reinvested in the future. As telomerase therapy comes into common medical use, it promises a growing population of healthy, independent elderly people and far more effective workforce.

THE ECONOMICS OF LONGEVITY

The key economic result of extending the human lifespan is that the workforce will become *more* productive, *more* energetic, and more efficient as we prevent aging disease and improve lifelong health. Telomerase therapy offers more productivity along with reduced costs for health care and elder care.

And yet, certain issues are unpredictable and raise questions. How high will the population densities climb? What will happen to the birth rate? To what extent will this increase environmental stress and related economic problems? Will we be able to adapt our legal structure, our retirement assumptions, and our social networks fast enough to cope with these problems? If we couldn't accurately predict how far we might extend the average lifespan— if prediction became impossible—how well would we cope with social changes "on the fly?" With greater uncertainty about life-spans, disease, and disability, we would run into financial instabili-ties that we had never encountered. How would we plan for our own futures, for the future of the company we work for, or for the future of our social systems when we could no longer base those plans on assumptions we have long taken entirely for granted?

If the individual outcome is an unalloyed good, the social out-comes are far less clear. Extending the healthy lifespan will not change terrorism, poverty, prejudice, or simple bad luck. While it can remove many of the ills of our own bodies, it cannot remove the ills of society. Even as we cure disease, the old curse remains: May you live in interesting times. Times will soon become more inter-esting than ever before. Of all the revolutions of human history— cognitive, agricultural, or industrial—the revolution caused by our ability to reverse aging may well be the most profound. Telomerase therapy offers the gift of life to individuals, but the gift of uncer-tainty to our society.

And yet, it is a hopeful uncertainty, even at its worst.

Compassion and Human Life

> Love cures people—both the ones who give it and the ones who receive it.
>
> — Karl Menninger

The objective of reversing aging is not to offer years, but compassion.

Our lives—human, social, familial, and personal—are not truly measured in years but in the quality of our experiences and the depths of our personal interactions. If life ends in misery, why would we extend it? But far more important, if your life is a pleasure to yourself and those who share it with you, why would we not extend the joy and delight of a healthy life?

Life is far more precious and deeper in meaning than the mere number of years we live.

Compassion for those around us is part of what makes our lives—and the lives of those to whom we extend our compassion—worth living. With modern medical care and in the midst of effective technology and competent clinical action, patients can still yearn for compassion. Just as we go to friends for a caring interaction, and not simply for solutions to problems, we go to physicians and hospitals for compassion, and not simply for diagnoses and therapy. This is not to minimize the value of medical knowledge, only to put it into perspective.

The secret of caring for patients is *caring* for patients.

A society that simply minimizes suffering with no attention to compassion is a society that has already failed. We don't live in a society merely hoping to avoid starvation and disease. We need to share in the lives and love of those around us. Merely to see society in economic or financial terms, or as a matter of population density or environmental concerns, is to miss the essence of human life. Compassion is as essential to society, to a healthy culture, as it is to good medical care.

What will happen to our culture and our lives when we reverse aging?

For many, the first questions are what can we afford, what about population, what about the environment? These issues need to be looked at honestly and carefully, but they are not the key issues in either our personal lives or in human culture. The key issue revolves around compassion, respect for ourselves, and our ability to dream and hope. The key difference between humans and other animals—and between us and our hominid predecessors—has been our ability to think about abstract things: compassion, respect, dreams, and hope among them. We can see the invisible, feel what we cannot touch, imagine what doesn't exist—this is what makes us human. But more important than our ability to envision things that don't exist is our ability to create such things, to make them real. We dream not to entertain ourselves, but to improve ourselves. Dreaming is valuable in itself, but to turn dream into reality is far more useful. Compassion is laudable, but we turn compassion from an emotion into an objective reality when we cure disease, when we give others the gift of long and worthwhile lives.

Which World Will We Choose?

Imagine a future in which, driven by population and economic issues, we provide medical care only to the young. At some age— perhaps seventy—we "put them out on the iceberg" simply because we regard those above that age as burdensome impediments to society.

Imagine another future in which the primary focus is on individuals and we have no age restriction on medical care. Whatever your age, the only question is whether we can help you, not whether you can help society.

Who among us would want to live in a world in which compassion is turned off because of age? What society can long survive when it urges death or suffering on certain members solely because they are older? Can any culture live when it blithely consents to death and disease?

We will soon have to answer these questions, and for different reasons than most of us have thought up to now. The new questions aren't about the morality and the cost of keeping elderly, often suffering people alive longer. They are about the morality of eliminating suffering and extending healthy, vital, productive life. How we make those choices will determine not only our own personal futures or what kind of culture we live in, but whether our culture survives with grace or perishes in ignorance. To reverse aging is not merely to extend lifespan, but to extend the human spirit. We have the opportunity to use that ability to prevent pain, fear, tragedy, and loss. How we help those around us, how we sculpt our laws and our society, will define who we are.

To use this opportunity with elegance, grace, and compassion is to succeed at being human.

Afterword

〜

As this book goes to press, the work moves ahead.

In early 2015, I founded Telocyte, a biotech company dedicated to the vision that forms the book you have just read. The project has an unprecedented understanding of aging and aging pathology, the skills to deliver our therapy, and a growing group of people—Peter Rayson, as well as Maria Blasco and her colleagues at the CNIO—who not only understand our scientific ability to cure Alzheimer's disease, but who are personally invested in that future, a future without Alzheimer's. Together, we are committed to taking our vision and ensuring that none of us need live in fear of aging and its diseases.

Impatient with theory, we are committed to a compassionate reality.

If you would like to help, feel free to contact us at Telocyte.com.

Glossary

Adenosine triphosphate (ATP). A coenzyme that transports chemical energy within cells for metabolism.

Adenovirus. A family of viruses that cause various degrees of minor upper respiratory illness in humans. Adenoviruses have long been a popular viral vector for gene therapy.

Anabolism. See *metabolism*.

Antioxidant. A substance that inhibits oxidative damage by free radicals and other oxidative molecules that damage cells.

Aplastic anemia. A disease in which the body no longer makes enough blood cells, including red cells, white cells, and platelets.

Apolipoprotein E4. A gene allele associated with elevated risk of various disorders, including atherosclerosis, Alzheimer's disease, ischemic cerebrovascular disease (stroke), and accelerated telomere shortening, among others.

Atrial fibrillation. An irregular heartbeat that increases the risk of stroke and other complications.

Base. In chemistry, a substance that can accept hydrogen ions, the opposite of an acid. In the context of this book, refers to any of the four nucleotides that make up the genetic code. (See *nucleotides*.)

Beta-amyloid. Peptides that form plaques as a result of aging microglial cells, resulting in neuron death and Alzheimer's disease.

Bisphosphonates. Drugs used to treat osteoporosis. They slow, but do not stop, bone loss.

Cardiomyocyte. Muscle cells that make up the cardiac (heart) muscles.

Carotid endarterectomy. A procedure in which narrowed carotid arteries are expanded by removing plaques, in hope of preventing strokes.

Catabolism. See *metabolism*.

Caudate nucleus. A nucleus in the base of the brain that is responsible for voluntary movement.

Cellulitis. A superficial bacterial infection of the skin, usually treated with oral antibiotics.

Cerebral cortex. The outer "gray matter" zone in the brain, made up of neurons that control movement, sensation, and other brain functions.

Chondrocytes. The cells that line the joints and make the joint smooth and friction-free, enabling normal movement.

Coxsackie. A common type of virus that causes a number of human viral infections, including viral meningitis.

C-reactive protein. A blood protein that indicates inflammation.

Cytokines. Small proteins that are secreted by cells to control other local cells.

Cytomegalovirus (human). A common virus that is present in most people and normally unnoticed, and which rarely causes significant disease in normal adults.

Cytotoxic cells. Cells of the immune system that are toxic to certain other cells, such as cancer cells.

Decubitus ulcers. Breakdown of the skin, usually found in elderly patients, when the body's weight presses on any area for a long time, resulting in death to the underlying tissue due to lack of blood supply. Also called bed sores.

Dopamine agonist. Used in hopes of alleviating Parkinson's symptoms. Parkinson's patients lose dopamine neurons, which use dopamine as a neurotransmitter.

Enzymes. Biological catalysts that accelerate chemical reactions in cells. Cells make three types of proteins: enzymes (which do all the work), structural proteins, and proteins that act as hormones.

Eosinophils. White blood cells responsible for combating parasites and certain other infections. (See *mast cells.*)

Farnesyltransferase inhibitors. Drugs used to limit the activity of the enzyme farnesyltransferase. They have been tried as a potential treatment for Hutchinson-Gilford progeria.

Fibroblast. One of the most common and widespread types of cell in the body. They make collagen, elastin, and other extracellular proteins. They also create other cells (such as fat cells), provide connective tissue, and repair tissue damage.

Free radical. An atom, molecule, or ion that has unpaired electrons. These make free radicals highly reactive, causing oxidative damage.

Glial cells. Non-neuronal cells that maintain, support (both metabolically and physically), and protect neurons (nerve cells) in the brain and peripheral nervous system. They have been implicated in causing Alzheimer's disease.

Hematopoietic cells. Cells (including stem cells) that produce all the various kinds of blood cells.

Homocysteine. An amino acid—high levels of which are associated with damage to endothelial cells, inflammation of the blood vessels, plaque formation, and resultant cardiovascular disease.

Hypercoagulation. Excessive blood clotting.

Inflammatory biomarkers. Substances found in the blood, elevated levels of which indicate various kinds of inflammatory diseases. (For example, see *C-reactive protein.*)

Insulin resistance. A problem typical of type 2 diabetes and in many elderly patients. Cells don't respond normally to insulin, even when present in normal levels.

Ischemia. Any time a tissue doesn't have enough blood supply to permit normal function. Ischemia is generally caused by problems with blood vessels. Heart attacks and strokes are examples of acute ischemia.

Isomerization. Many complex molecules can be folded into different structures, even with exactly the same chemical structure. This can occur spontaneously, even at normal body temperature, and often results in molecules that are no longer functional.

Keratinocytes. The most common cell type in the epidermis, these cells create the outer layer of skin, which typically sloughs off and is continuously being replaced by cells in the lowermost layer of the epidermis.

Leptin. A hormone that controls fat deposition and inhibits hunger.

Leukocytes. White blood cells. These are the primary cells of the immune system and circulate throughout the body.

Liposome. A tiny artificial "bag" made of lipid molecules, used to deliver drugs.

Lymphokine. A type of cytokine made by lymphocytes to control the immune system functions.

Mast cells. A special immune cell often involved in allergies and inflammation.

Metabolism. The chemical reactions in cells that supply energy to the cell, as well as creating and breaking down biological molecules. Metabolism has two parts: Anabolism is the creation of molecules; catabolism is the breaking down of molecules.

Methylation. An alteration of DNA used to control gene expression. Epigenetic changes frequently rely on methylation and similar changes.

Microglia. A glial cell, much like a macrophage, found in the nervous system. The aging of this cell results in Alzheimer's disease.

Nucleotides. The collective term for the four types of molecule that make up the language of the genetic code in our DNA: adenine, guanine, thymine, and cytosine.

Oxidants. Chemical agents that oxidize other molecules by taking away electrons. An example would be oxygen combining with iron to form rust. In human physiology, free radicals cause oxidative damage to cells. (See *free radicals*.)

Peristaltic waves, peristalsis. The process of muscular contractions by which food, and eventually food waste, is moved through the digestive tract from the esophagus to the stomach, and subsequently through the bowels.

Pluripotent stem cells. Stem cells that have the potential to differentiate into any type of cell found in the human body.

Proteoglycans. Complex substances—partly protein, partly complex sugars—making up part of the extracellular matrix found between cells.

Prothrombotic mutations. Mutations that cause excessive blood clotting.

Restenosis. A recurrence of stenosis—the re-narrowing of an artery after treatment to clear blockage. (See *stenosis*.)

Resveratrol. A common plant substance found in grapes, blueberries, raspberries, and mulberries. Although it has been touted as beneficial for treating heart disease and cancer, boosting metabolism and anti-aging, there is limited evidence of these health effects in humans.

Senescence. Biological aging. This term is often used in regard to cells (as opposed to organisms).

Somatic cells. Any cell forming the body of an organism, as opposed to the sexual cells responsible for reproduction (i.e., sperm and ova).

Southern blot. A process used to separate, detect, and measure biological molecules in the lab, including proteins and DNA.

Stenosis. A narrowing of a blood vessel, leading to restricted blood flow.

Substantia nigra. Deep brain nucleus that controls movement and is usually damaged in patients with Parkinson's disease.

Synovial fluid. A viscous fluid found in joint spaces (e.g., knees, hips, ankles, wrists, elbows, shoulders) that reduces friction between the articular surfaces during movement.

Tau proteins. Proteins that are abundant in many neurons. Abnormal tau proteins (tau tangles) are often found in patients with Alzheimer's disease.

Telomeres. DNA structures at the ends of chromosomes that shorten with each cell division.

Thymus. A special immune system organ that is the source of T cells, which are part of the adaptive immune system.

Tocopherols. Fat-soluble compounds with vitamin E activity. Tocopherols are a group of compounds that can be collectively referred to as vitamin E.

Umami. One of our five basic tastes (the others are sweet, sour, bitter, and salty). People often describe umami as having a "brothy" or "meaty" taste.

Viral vector. A virus used to deliver a therapeutic molecule, such as a gene. The internal part of the virus is generally removed and replaced by a drug or therapeutic gene, while the external shell of the virus is useful in ensuring that the drug or gene can be delivered into the target cells.

Index

Acknowledgments

No writer is an island. My ideas—correct or not—are my own, but the final book is the result of those who believed in their own data (Len Hayflick and Maria Blasco), believed in me (my wife, Joy), or believed in this book (Glenn Yeffeth and Dave Bessmer).

The work that grows from all of this, the attempt to cure human disease, is equally due to those who believed in the prospects, especially Maria Blasco, but also Peter Rayson and Brad Edwards. For the sake of those who depend on us, may we all be proven right.

The world depends not on what we think or what we write, but on what we prove.

About the Author

⚓

D r. **Michael Fossel** earned his PhD and MD from Stanford University, where he taught neurobiology and research methods. Winner of a National Science Foundation fellowship, he was a clinical professor of medicine for almost three decades, the executive director of the American Aging Association, and the founding editor of *Rejuvenation Research*. In 1996, he wrote the first book on the telomere theory of aging, *Reversing Human Aging*, describing the prospects for extending telomeres, reversing aging, and curing age-related disease. Author of more than sixty scientific articles, he authored the sole medical textbook in the area of telomeres and clinical therapy, *Cells, Aging, and Human Disease* (Oxford University Press, 2004). He is highly regarded for his remarkable university course, the Biology of Aging.

The world's foremost expert on the clinical use of telomerase for age-related diseases, he has lectured at the National Institutes of Health and the Smithsonian Institution, and still lectures at universities, institutes, and conferences throughout the world. He has appeared on *Good Morning America*, ABC *20/20*, NBC *Extra*, Fox Network, CNN, BBC, Discovery Channel, and NPR.

He is currently working to bring telomerase to human trials for Alzheimer's disease.

His website is www.michaelfossel.com.